URBAN FOOD PRODUCTION FOR ECOSOCIALISM

This book explores the critical role of urban food production in strengthening communities and in building ecosocialism. It integrates theory and practice, drawing on several local case studies from seven countries across four continents: China, Cuba, Ghana, Italy, Tanzania, the UK, and the US.

Research shows that the term "urban agriculture" overstates the limited food-growing potential in cities due to a shortage of land required for growing grains, the basic human food staple. For this reason, the book suggests "urban cultivation" as an appropriate term which indicates social and political progress achieved through combined labours of urbanites to produce food. It examines how these collaborative food-growing efforts help raise local social capital, foster community organisation, and create ecological awareness in order to promote urban food production while also ensuring environmental sustainability. This book illustrates how urban cultivation constitutes a potentially important aspect of urban ecosystems, as well as offers solutions to current environmental problems. It recentres attention to the global South and debunks Eurocentric narratives, challenging capitalist commercial food-growing regimes and encouraging ecosocialist food-growing practices.

Written in an accessible style, this book is recommended reading about an emergent issue which will interest students and scholars of environmental studies, geography, sociology, urban studies, politics, and economics.

Salvatore Engel-Di Mauro is Professor in the Department of Geography at SUNY New Paltz. Research interests include critical physical geography, socialist movements, soil degradation, urban soils, and trace element contamination. He is the chief editor for the journal *Capitalism Nature Socialism*.

George Martin is Emeritus Professor, Montclair State University and Visiting Professor, Centre for Environment & Sustainability, University of Surrey. He is a senior research fellow at the Center for Political Ecology, University of California, Santa Cruz, and advisory board member for *Capitalism Nature Socialism* at Routledge. His focus is urban environmental sociology. Professor Martin's recent publication was *Sustainability Prospects for Autonomous Vehicles* with Routledge in 2019.

URBAN FOOD PRODUCTION FOR ECOSOCIALISM

Cultivating the City

Salvatore Engel-Di Mauro and George Martin

Routledge
Taylor & Francis Group
LONDON AND NEW YORK

First published 2022
by Routledge
2 Park Square, Milton Park, Abingdon, Oxon OX14 4RN

and by Routledge
605 Third Avenue, New York, NY 10158

Routledge is an imprint of the Taylor & Francis Group, an informa business

British Library Cataloguing-in-Publication Data
A catalogue record for this book is available from the British Library

Library of Congress Cataloging-in-Publication Data
A catalog record for this book has been requested

ISBN: 978-0-367-67417-5 (hbk)
ISBN: 978-0-367-67418-2 (pbk)
ISBN: 978-1-003-13128-1 (ebk)

DOI: 10.4324/9781003131281

Typeset in Bembo
by Apex CoVantage, LLC

CONTENTS

FIGURES

TABLES

FOREWORD

Clarification on meaning of key terms

Central to this book is the repeated use of terms to categorise materials we use in our analyses. They are general and fluid rather than specific and fixed and thus merit some explication. They are united here in their relevance to our ecosocialist perspective on urban food production. The terms comprise larger domains than those we address.

The first is a melange of food-focused idioms: ***Food security, food access and food sovereignty***. Security and access are linked in usage and speak to the supply and distribution of food—its availability and sufficiency (FAO 2002; WFP 2018). They feature in the discourses of governments and non-governmental organisations, including the United Nations (WESP 2014), the World Bank, and the OECD (2007). The terms originated in the 1970s with national poverty and famines as their reference points. Sovereignty appeared in the 1990s and develops a politics of food as an issue of local control and as a human right, with emphasis on the equity and equality of entire food systems (GJN 2018; Patel 2009; Roman-Alcalá 2018). It has become a defining principle in food justice political movements. Food sovereignty is based in a wealth of organisations and associations, one of which is *Via Campesina*, a global alliance of local and national peasant rights groups that was founded in 1993. The term is used in both the security and sovereignty senses here, essentially with security as relevant to ecology issues and sovereignty to socialist ones.

The second set of terms is categorical—several various sorting frames for the world's nations according to their comparative economic, geographical, historical, and political characteristics. This idiom has its roots in the Cold War's East–West division, which expanded to include the entire globe with the First (West)—Second (East)—Third (non-aligned) trio. The latter was often used to refer to post-colonial (formally independent) nations in Africa, Asia, and Latin America. The trio slowly disappeared after the capitalist transformation of Central and Eastern European socialist states in 1989. The research and academic communities continued the categorisation—after all, the traditional first task of scientific endeavour is identification and categorisation. The new conventional nomenclature focused on the economic status of nations and was based in data on income. It is another trio—of *less developed, developing, and more developed countries* (sometimes also featuring least developed). The common measure used is Gross National Income, and its users include the United Nations (UNSD 2018), International Monetary Fund, World Bank, and Organisation for Economic Cooperation and Development. The categorisation is described as meant for comparative statistical and analytical convenience and is not supposed to represent a judgement about nations. But, of course, it does have frequent social, political, and cultural applications.

A parallel alternative set of terms, emerging by the 1970s and associated especially with world-systems frameworks, describes countries in relation to their position in the capitalist world economy, whose main institutional pillars are national states (Amin 1977). Accordingly, most countries exist in conditions of economic dependence on more powerful semi-peripheral countries and even more on the most powerful, core capitalist countries. [Kwame Nkrumah (1965) identified this situation as neo-colonialism, which seems often a much more accurate designation.] Though we find broad agreement with this sort of perspective and much of what we write implies in part a specifically Marxian world-systems approach, we refrain from using such terminology on account of its weaker resonance among activists.

Since the 1990s, a different set of terms has become of much wider use. Dislike of the implicit (capitalist) political bias in use of the word "development" became an impetus for popularising an equator-based duo—*global North and global South.* It is similar in meaning to the traditional distinction of rich and poor nations but with political shading added. The global North contains the most powerful countries of the world (the core countries in world-systems perspective), based mostly in North America and Western Europe (the former First World), and increasingly some nations of East Asia and Oceania (e.g. Australia, Japan, New Zealand), while the global South is comprised of the less powerful nations in Latin America, Africa, the Asia (the former Third World). All the categories are rather arbitrary and challenged by exceptions, such as Australia and New Zealand being placed in the global North (IPI 2010). They are used in rather stereotypical and shorthand fashion. However, they are useful and even necessary in the analysis of many subjects, including food. Both the development and equatorial terminologies are used here, primarily to be consistent with cited sources and wider political movements beyond academic circles.

Such categories are supplemented by a third, much less frequently encountered but more precise set of terms, ecologically rich and ecologically indebted, that we use especially when addressing the ecological and physical aspects of cities. These refer to national levels of capital accumulation and ecological impact and status, such as can be assessed using Ecological Footprint estimates (see Frey 2012; Salleh 2009). Using these parameters enables us to highlight the fact that capital-rich countries (the global North) are typically ecologically impoverished or indebted to the rest of the world. Another way of putting this is that wealth levels (of capital accumulation) in global North countries are predicated on taking resources from and thereby undermining or destroying ecosystems in the rest of the world. The opposite tends to be true in the global South, where there is high ecological "wealth" but low capital accumulation.

The fourth term is the concept of *social capital*. It is meant to describe aspects and products of social life that become ongoing assets for individuals and groups (Gauntlett 2011; HKS 2006). Specific examples include mutual aid, networking, and education, all of which may contribute to collective action for social change. Its first use may have been in a 1916 book by Lyda Hanifan on how neighbours could cooperate to oversee schools. It was developed in the 1980s by Bourdieu (1986) and Coleman (1988) and later popularised by Putnam (2001). Social capital is *de novo* apolitical—it can be exclusionary as well as inclusionary. It can be used to defend elitism as well as to attack it. Its usefulness centres on the fact that it counters the individualistic rational choice model of human behaviour prevalent in capitalism and its political philosophy excrescences e.g. liberal democracy. Social capital is a stock of acquired knowledge and information, often informally sourced and free of charge, that is, acquired outside capitalist processes and then utilised for collective purposes. It is used here to describe the promising contributions that urban food growing can make to social sustainability, especially through environmental education that promotes an understanding of ecologies that can be a basis for collective action and social change. Furthermore, drawing from Karl Marx, social capital to us also implies one of the panoply of tools for the eventual full development of human potentials to overcome the different kinds of alienation resulting from capitalist relations—alienation from the products of one's labour, from the production process, from nature, and from each other (Marx 1844, 70–81; see also Mobasser 1987).

A fifth set of inter-related terms is *ecological, (physical) environmental, and biophysical*. This distinction will not be made apparent until Chapter 4, just to make it easier for the text to flow. But after that, we use biophysical to refer to both ecosystems and physical environments because organisms are not the same as the physical environments that they (and we) may inhabit. We draw from ecologists in our use of the word ecological as relations among organisms and between organisms and physical processes (Haila and Levins 1992, ix). Physical processes include forces, like wind or water flow, or environments, like glaciers or soils (a force can also be viewed as a physical environment and vice versa, depending on the research question). Ecosystems and physical environments overlap because ecosystems include abiotic components (i.e. physical environments), like rivers, minerals, and air humidity.

The final terms refer to the perspectives that inform our work: *political ecology, ecofeminism*, and *ecosocial relations*. The political ecology idiom emerged in the 1980s as a hybrid of anthropology, development studies, ecology, geography, and political science (Brannstrom 2013; Gezon and Paulson 2005; Khan 2013; Robbins 2019). Since then it has fragmented into active research communities, including those who borrow from environmental justice and sustainability, which is the sense in which it is primarily used here. Our ecosocialist, egalitarian, and sustainability perspective utilises the work of the idiom's Marxist approaches. This is coupled with *ecofeminist* perspectives, where multiple forms of oppression are viewed as material conditions inseparable from and causal of ecological destruction. It is a worldview that originated in the feminist movement during the 1970s and has diversified into many ways of understanding the nexus of oppression and environmental degradation, ranging from linkages to patriarchal heteronormative racial ideologies to connecting those ideologies and associated oppressive practices to human supremacism and speciesism, as they lead to exploitation and mass slaughter of other organisms and to life's sixth mass extinction (D'Eaubonne 1974, 221; Mies and Shiva 2014; Plumwood 1993; Salleh 1997; Turner and Brownhill 2006).

The other foundational term, *ecosocial relations* (Engel-Di Mauro 2014), is a dialectical historical materialist perspective we use that is inspired by Marxists working in the biophysical sciences (especially Levins and Lewontin 1985). Simply stated, historical materialism means focusing on changes over time (history); however, we emphasise ecosystems as well because society is part of and a constituent of ecosystems. Ours is a materialist approach in that people's and other organisms' changing conditions are the motor of history, including changes to our ideas about the world. In this framework, historical material change happens dialectically. For our purposes here, biophysical dynamics and social relations are mutually influential and mutually transformative, that is, people change themselves in the process of impacting the rest of nature (or biophysical conditions). Our approach is, overall, an application of existing frameworks. We make no claims to theory development. The task is instead to analyse and critique current theoretical approaches as they are applied to urban food growing and to offer recommendations regarding urban food production's usefulness in building an ecosocialist future.

References

Amin, S. 1977. *Lo Sviluppo Ineguale. Saggio sulle Formazioni Sociali del Capitalismo Periferico* [Unequal Development. Essay on the Social Formations of Peripheral Capitalism]. Torino: Einaudi.

Bourdieu, P. 1986. The Forms of Capital. In, *Handbook of Theory and Research for the Sociology of Education*, edited by J. Richardson, 241–58. Westport CT: Greenwood.

Brannstrom, C. 2013. *Political Ecology*. Oxford: Oxford Bibliographies, Oxford University. www.oxfordbibliogrphies.com/view/document/obo-978019987/obo-9780199874002-0002.xml (Accessed 13 August 2020).

Coleman, J.S. 1988. Social Capital in the Creation of Human Capital. *American Journal of Sociology* 94: S95–S120.

d'Eaubonne, F. 1974. *Le Féminisme ou la Mort* [Feminism or Death]. Paris: Horay.

Engel-Di Mauro, S. 2014. *Ecology, Soils, and the Left: An Ecosocial Approach*. New York: Palgrave McMillan.

FAO. 2002. *Food Security: Concepts and Measurements*. Rome: Food and Agriculture Organization, United Nations. www.fao.org/docrep/005/y4671e/y467e06.htm (Accessed 13 August 2020).

Frey, R.S. 2012. The e-Waste Stream in the World-System. *Journal of Globalization Studies* 3: 79–94.

Gauntlett, D. 2011. *Making Is Connecting: The Social Meaning of Creativity*. Cambridge: Polity Press.

Gezon, L.L., and S. Paulson. 2005. Place, Power, Difference: Multiscale Research at the Dawn of the Twenty-first Century. In, *Political Ecology Across Spaces, Scales, and Social Groups*, edited by S. Paulson, and L. Gezon, 1–16. New Brunswick: Rutgers University Press.

GJN. 2018. *The Six Pillars of Food Sovereignty*. London: Global Justice Now. www.globaljustice.org.uk/six-pillars-food-sovereignty.

Haila, Y., and R. Levins. 1992. *Humanity and Nature: Ecology, Science, and Society*. London: Pluto Press.

Hanifan, L.J. 1916. The Rural School Community Center. *Annals of the American Academy of Political and Social Science* 67: 130–8.

HKS. 2006. *Social Capital Theory*. Cambridge MA: Kennedy School, Harvard University. https://sites.hks.harvard.edu/saguaro/web%20docs/GarsonSK06sullabus.htm (Accessed 13 August 2020).

IPI. 2010. *Deconstructing the North-South Label*. New York: International Peace Institute. www.ipinst.org/2010/04/deconstructing-the-north-south-label (Accessed 13 August 2020).

Khan, M.T. 2013. Theoretical Frameworks in Political Ecology and Participatory Nature/Forest Conservation: The Necessity for a Heterodox Approach and the *Critical Moment*. *Journal of Political Ecology* 20: 460–72.

Levins, R., and R. Lewontin. 1985. *The Dialectical Biologist*. Cambridge: Harvard University Press.

Marx, K. 1844 [1978]. The Economic and Political Manuscripts of 1844. In, *The Marx-Engels Reader*, edited by R.C. Tucker, translated by M. Milligan, 2nd ed., 66–135. New York: W.W. Norton.

Mies, M., and V. Shiva. 2014. *Ecofeminism*, 2nd ed. London: Zed Books.

Mobasser, N. 1987. Marx and Self-Realisation. *New Left Review* 161: 119–28.

Nkrumah, K. 1965. *Neo-Colonialism, the Last Stage of Capitalism*. London: Thomas Nelson & Sons.

OECD. 2007. *Insights: What Is Social Capital?* Paris: Organisation for Economic Cooperation and Development. www.oecd.org/insights/37966934.pdf (Accessed 13 August 2020).

Patel, R. 2009. Food sovereignty. *The Journal of Peasant Studies* 36: 663–706.

Plumwood, V. 1993. *Feminism and the Mastery of Nature*. London: Routledge.

Putnam, R.D. 2001. Social Capital: Measurement and Consequences. *Isuma: Canadian Journal of Policy Research* 2: 41–51.

Robbins, P. 2019. *Political Ecology: A Critical Introduction*, 3rd ed. New York: Wiley.

Roman-Alcalá, A. 2018. (Relative) Autonomism, Policy Currents and the Politics of Mobilisation for Food Sovereignty in the United States: The Case of Occupy the Farm. *Local Environment* 23: 619–34.

Salleh, A. 1997. *Ecofeminism as Politics. Nature, Marx, and the Postmodern*. London: Zed Books.

Salleh, A., ed. 2009. *Eco-Sufficiency and Global Justice: Women Write Political Ecology*. London: Pluto Press.

Turner, T., and L. Brownhill. 2006. Ecofeminism as Gendered, Ethnicized Class Struggle: A Rejoinder to Stuart Rosewarne. *Capitalism Nature Socialism* 17: 87–95.

UNSD. 2018. *Methodology*. New York: United Nations Statistics Division. https://unstats.un.org/unsd/methodology/m49/ (Accessed 13 August 2020).

WESP. 2014. *Country Classification: Data Sources, Country Classifications and Aggregation Methodology*. New York: World Economic Situation and Prospects, Department of Economic and Social Affairs, United Nations.

WFP. 2018. *What Is Food Security?* Rome: World Food Programme, Food and Agriculture Organisation, United Nations.

ABBREVIATIONS

BLM	Black Lives Matter
BP	before present
C	centigrade/Celsius
ca	circa
CAFO	concentrated animal feeding operations
CDC	Centers for Disease Control and Prevention (US)
cm	centimetre
CO_2	carbon dioxide
COVID	COronaVIrus Disease
DDT	dichloro-diphenyl trichloroethane
EU	European Union
FAO	Food and Agriculture Organisation (UN)
GHG	greenhouse gases
GIS	geographic information system
GMO	genetically modified organism
GDP	gross domestic product
ha	hectare
kg	kilogramme
km	kilometre
lb	pound
LCA	life cycle assessment
m	metre
mg	milligram
mi	mile
mm	millimetre
NO_2	nitrogen oxide
PAH	polycyclic aromatic hydrocarbon

PCB	polychlorinated biphenyl
pH	potential of hydrogen (measurement of acidity)
PM	particulate matter
ppm	parts per million
PRC	People's Republic of China
UA	urban agriculture
UK	United Kingdom
μm	micron
UN	United Nations
US	United States
USSR	Union of Soviet Socialist Republics
WHO	World Health Organization (UN)
WSCG	West Side Community Garden

1

URBAN AGRICULTURE AND ECOSOCIALISM

Growing food in cities is commonly referred to in academic and expert discourse as urban agriculture, including in the FAO (2018) definition. The term combines two Latin root words—*ager* for field (from the Greek *agros*) and *cultura* for growing or cultivation—and field crops are its common denominator worldwide. Using the word agriculture to describe urban food growing of ecologically indebted countries projects a level of sustainable food production that is not physically achievable—not now and most likely not in the foreseeable future (Martin, Clift, and Christie 2016), irrespective of socially insensitive techno-fantasies such as vertical farming (see, e.g. Despommier 2018). Agricultural-scale food production is not compatible with urban landscapes. One estimate is that meeting just the vegetable consumption of urbanites would require about one-third of the world's total urban area (Martellozzo et al. 2014). This estimate is generous in that it does not take into consideration how much urban land is contaminated or otherwise unavailable to grow food. Converting this much urban land to growing vegetables is not a reasonable prospect.

Urban areas are where most food consumers now dwell. Cities do not contain the expanses of ground-level land fully exposed to the sun that is needed for field crops, largely the cereal grains that comprise the bulk of food most of us consume. Agricultural fields are primarily used to grow single crops that provide this "staff of life." The ecologically and socially destructive character of this and other aspects of profit-oriented industrial food production are highlighted here and in the next chapter—in short, growing food in cities cannot replace conventional agriculture, much less undo its long-term harm. In 2001, of the world's agricultural land in crops, 44 per cent was under cereal cultivation and only 6 per cent in fruit, vegetable, and melon cultivation (FAO 2004). Even in our contemporary multi-diet world, cereal grains, led by maize, rice, and wheat, globally account for about one-half of people's dietary intake (FAO 2016). The remainder comprises a wide

DOI: 10.4324/9781003131281-1

variety of plants—palm oil, coffee beans, tea leaves, etc. The production of meat uses even larger plots of land. Even sustainably produced meat requires flocks and herds of animals that are pastured in sizeable outdoor expanses—70 per cent of all global agricultural land is devoted to animal grazing. In addition to the land required, there are major public health concerns associated with groups of animals being farmed in proximity to urban residents as well as pandemic and other risks in expanding agriculture and urban areas into tropical forests.

Because of a lack of land, urban food growing can sustainably produce for only a relatively small portion of urban diets, consisting of fruits, vegetables, and leaves. Even this small niche of food growing is limited in cities of the global North by the contamination of lands abandoned in their de-industrialisation. Moreover, small plots present scaling obstacles to growing food more sustainably and more efficiently. So, instead of viewing urban food growing as primary, it is more realistic to see it as a secondary gain. An alternative designation—cultivation (see Martin, Clift, and Christie 2016; WinklerPrins 2017)—is a more accurate definition upon which to base any related analysis, including what we would advocate for, an ecosocialist programme. Cultivation is a word that connotes a range of commonly used etymological meanings—to cultivate is to learn, to develop, to be urbane, as well as to work the soil. From an ecosocialist perspective, urban food cultivation can be a vehicle for an integration of overlapping biophysical and social spheres. It has, as its mobilising strategy, been advocating for the hands-on experience of growing food to promote ecological learning as well as community organising for food justice. The point is not that urban food growing is ecologically and socially unimportant; rather, its primary importance does not lie in being agriculture. As is often the case in human endeavours, unachievable goals can result in secondary benefits.

The historical changes to food procurement systems over the past several centuries result from an increasing concentration of land control, a direct consequence of capitalist development. Rural clearances linked to urban growth are not just a feature of ecologically indebted countries and their distant pasts. They are present today in many countries—Brazil, China, India, and Palestine, to list a few. Countries like Italy and the US are hardly immune, as their remaining farms continue to be corporatised, attracting migrants from elsewhere who have been dispossessed from the land in their home countries. Such clearances are far from complete, and many people continue to resist their eviction. Sometimes they forge ways of pre-empting land expropriation or of re-appropriating land, for example, by squatting and land occupations.

The historical development of food–city distanciation has become a worldwide affair. The global economy, in which the food industry is one of the largest sectors, is dominated by neoliberal capitalism which crosses and links North–South distinctions. For example, rural areas of the global South are home to plantation-scale production of agricultural commodities consumed in cities of the global North, directly in diets that include fruit such as bananas and indirectly in food processing with ingredients like palm oil. In cities of the global North, the present renewal of urban food growing has been generated by several factors, including the in-migration of displaced farmers from the global South. However, cities in the global South are the primary destinations of their country's landless peasant farmers. China and India are the foremost examples, as both are in the midst of the largest rural-to-urban population movements in human history (UN 2014). Other nations in the global South are on the same path, including Vietnam (Kurfurst 2019).

A principal outcome of uneven development fostered by capitalist colonial policies is that the world is characterised by high and still increasing food inequalities, within as well as between countries (Ponting 2007, 198). Uneven development means that improvements in material well-being enjoyed by some in some places is the result of exploitation and resource extraction and the consequent reduction in the material well-being of the many in other places (Amin 1974; Frank 1966; Rodney 1974). This is not just between countries or imperial centre and colonies. It is a process that happens at any scale, from a neighbourhood to the world. This is among the main reasons why the level of food consumption (but also production) varies wildly around the world. The relative cost of a simple plate of pulses and vegetables differs markedly between countries in the global North and those in the South (WFP 2017). In Sweden, food purchases represent only 13 per cent of daily expenses; in Tanzania, they represent 73 per cent; and in Bangladesh, 56 per cent (Southgate, Graham, and Tweeten 2007, 31). Food inequalities also exist within countries. Thus, within the US about twice as many persons with low incomes have poor diets as compared to persons with the highest incomes (Rehm et al. 2016).

The development of large-scale industrial farming is exemplified by the US, and one of its most conspicuous present venues is concentrated animal feeding operations (CAFOs). They are farms in which a minimum of 1,000 "animal units" are confined in grass-free feedlots for at least 45 days to be fattened up before butchering (USDA 2017). They discharge their voluminous waste into manure lagoons. Because animal waste carries potentially harmful pollutants for human health and ecosystems, they are regulated by the US Environmental Protection Agency. Through the extensive web of inter-linkages in global neoliberal capitalism, countries in the South have been catching up to the US in building feedlots (Harvey et al. 2017). The CAFOs now account for most of the world's poultry and pork production (UN 2016).

The global South is home to 80 per cent of the world's urban food growing. The Resource Center for Urban Agriculture and Forestry reported that Hanoi and Havana are leading sites (Rose 2016, 173–4). Whereas cities of the North grow for niche markets such as salad greens destined for restaurant tables, many cities of the South produce about one-third of the total food consumed by their residents (Mok et al. 2014; Orsini et al. 2013; Thebo, Dreschel, and Lambin 2014; Zezza and Tasciotti 2010). However, the situation is not as easily defined as this. There are cities in the global South, for instance Chongqing, where food production may be sizable, but others, like Shanghai or Beijing, where it is possibly less substantive than even in cities of the global North (Cai and Zhang 2000; Peng et al. 2015; Peng and Hu 2015). There is also a wide variety and scale of urban food growing in parts of East and Southern Africa, promoted by a combination of extreme rural poverty and rural-to-urban migration. For example, most of Lusaka's residents grow their own food, often on unused land, while in other cities it is illegal due to health concerns, especially when keeping animals is involved (Beach 2013).

A critical perspective on the study of urban agriculture

In both the global North and South, or rather, between the ecologically indebted and endowed (but ransacked), the major overlaps in the varieties of projects which fall under the umbrella of urban food production are being increasingly appreciated, and the approaches predicated on their division are being questioned (WinklerPrins 2017). The division masks a contradictory and narrow view of the world that universalises urban processes based on European and associated settler colonial contexts, and simultaneously treats the global South as inherently different and unrelated to the pre-eminence of the global North. Eurocentrism is also among the factors that have contributed to making many leftist approaches parochial enterprises prone to universalising what is specific to European histories. Continuing with such a mindset risks scuppering potentials for creating cross-cutting, internationalist progressive political alliances. The global inter-linkages in present-day food production and consumption call into question the ecological sustainability of the limited scale of growing food in cities.

Yet most who work within the social sciences, particularly in urban policy and planning, design, and architecture, appear not to concern themselves with the ecological basis of city food growing at all. A general and persisting lack of attentiveness to biophysical processes is shared by conventional and critical approaches alike. It is exacerbated by ignoring or misunderstanding contamination processes, as if resolving them was a relatively simple technical matter—such as using raised beds or growing vegetables vertically or on roofs (Gorgolewski, Kommisar, and Nasr 2011, 14; Reynolds and Cohen 2016, 3). Generally, there is little sense of how biophysical processes and pollution legacies from industrialisation affect urban food-growing projects and urban politics more widely. It is also the case that environmental scientists themselves tend to be remiss in grasping and incorporating social and political considerations; rather, they often lose themselves in generic lists of variables that are often treated as if causally disconnected from each other (e.g. pollution, population growth, wealth levels, and urbanisation). The result is that social differences and political struggles related to urban food growing are folded into a homogenised mass and treated as if they are unrelated to environmental problems such as ongoing urban pollution and land-use conflict.

Thankfully, we are not starting from scratch. Much work has long been done that is attentive to relations of power as well as critical of capitalist relations. The social justice prospect of growing food in cities is finally becoming a basis for political discussions and mobilisations, but so far, curiously, not the environmental justice component. Reynolds and Cohen (2016), for example, employ critical race theory to expose the racialised and classist basis of much urban food-growing discourse. Along with others, they demonstrate that growing food is often an unrecognised form of everyday political resistance (Dawson and Morales 2016; Eizenberg 2013). Some argue that agroecology-based urban food production stimulates the development of notions of incommensurability (i.e. things are not always exchangeable), promotes more caring (for others, other species, etc.), and enables more local resourcefulness, all of which go against capitalist logic (Tornaghi and Dehaene 2020). Such findings and interpretations of existing urban food production form part of an ongoing re-framing of cities as sites for political mobilisations involving issues of environmental and food justice (Heynen, Kaika, and Swyngedouw 2006; Loftus 2012; Pulido 2000).

The trouble for us is that existing political ecological and other similar critiques of urban food growing are rarely, if at all, complemented by explicit politics (save, at times, for allusions to some unspecified "radicalism"). Just as problematic is that these kinds of alternative perspectives are insufficiently (and often not in the least) mindful of the physical and ecological processes affecting the very feasibility of urban food production—under any political regime and relative to any political project. More disconcerting is that such critiques are often so vague about specifics as to be applicable to a wide variety of dissimilar political ends. This is because they fail to develop explicit political platforms within which urban food cultivation can be situated. Here, McClintock's (2014) warning is apt regarding the need for contextualisation and appreciation of the variety of political contradictions ensconced in urban food growing. Or, as in Pollan's (2016, 81) words, the urban food movement is "a collection of disparate groups that seek change but don't always agree with one another on priorities."

Addressing the social and biophysical issues in, and challenges of, urban food production means looking into both the biophysical and social aspects of cultivation. This is necessary to envisage healthier and more egalitarian alternative futures. The timing seems ripe for this. In academic settings and technocratic planning circles, a combined environmental and social approach is increasingly viewed favourably as contributing to developing the transdisciplinarity beckoned by the finally perceived need and promotion of urban ecological sustainability (Bell et al. 2016; Bernstein 2015; Gorgolewski, Kommisar, and Nasr 2011). However, this is not what motivates us to cross diverse fields of knowledge. Research should not be reduced to purely academic or technocratic prerogatives.

We are not interested in limiting our work to discovering and trying to explain. We are not terribly keen on counterbalancing the erasure of relations of power and social inequalities with critiques or alternative understandings of the way things are, as important as that line of work is. We instead reckon urban food production should be evaluated according to how it fits or can be made to fit broader political objectives instead of trying to prove how, say, egalitarian anti-capitalist politics can be relevant to or emergent in urban food production. In any case, the vogue of hiding behind hazy notions of emancipation or of more just worlds is no counterweight to the business-friendly and brazenly pro-capitalist scholarship pervading many writings on the topic. Worse, at a time when socialist ideas are regaining a serious hearing in the demographically tiny global North (unlike in the rest of the world, where they never went away), much self-styled critical scholarship on "political gardening" studiously avoids socialism, save for re-dredged bygone "Cold War" caricatures, groping for some fuzzy "radicalism" or "progressivism" for the global North's all too often well-to-do, white-dominated urban gardening communities. [See, for instance, the collection edited by Tornaghi and Certomà (2019), where there is only one positive reference to socialism and indirectly at that, by way of a Lefebvre-inspired Marxist take from Purcell and Tyman (2015).]

This is why we are more interested in formulating ideas useful towards the development of an explicit political project, and one that is gaining momentum especially in the global South: ecosocialism (or its equivalent, under whichever label people, especially organisers, deem appropriate). As will be clearer below, our understanding of ecosocialism departs from state-centred forms of socialism but is not necessarily inimical to them. This is because, to put it concisely, ecosocial contexts and histories matter, so it makes no sense to expect that political strategies that work in one place can be plopped seamlessly elsewhere. Differing place-specific strategies, though, should be ideally coordinated at the global scale, or at least not undercut each other. In such an ecosocialist perspective, urban food cultivation is one among many related activities that can lead to the development of ecosocialist ways of life. In other words, urban food cultivation needs to be evaluated according to how well it can cohere with wider ecosocialist struggles. We think what is needed is a view that considers both social justice (e.g. food justice and sovereignty as well as environmental justice) and ecological sustainability, which implies environmental justice. This is otherwise known as a red-green perspective, which we ground in an ecosocial framework (Engel-Di Mauro 2014).

Such a perspective draws from a variety of approaches. For those focused mainly on the social power relations, these include eco-anarchism (e.g. squatters' movement perspectives), materialist ecofeminism, eco-Marxism, environmental and food justice, and urban political ecology. However, other approaches are related to more general urban processes. Principal among these are the materialist ecofeminists' subsistence and eco-sufficiency perspectives (Mies and Bennholdt-Thomsen 1999; Salleh 2009), whereby urban food production is evaluated relative to its contribution to the development of a society in which satisfaction of peoples' basic needs prevails without undermining ecological and social conditions (O'Connor 1998). Another more general approach is Lefebvre's (1974) analysis of urban space as socially produced and his conceptualisation of the city as simultaneously material, representational, and lived. All three vectors are relevant to constituting an ecosocial platform for urban food cultivation. Relevant to this platform is Eizenberg's (2013) use of Lefebvre's framework to analyse the reformist co-optation of community gardens by city governments. Lefebvre's subsequent reformulation of urbanisation as a global process (Madden 2012) coheres with our multiple-scaled approach which is attentive to inter-linkages as they have been developed in political ecology. This approach overlaps with the world-systems paradigm that initially was a foundation of political ecology (Blaikie 1985; Engel-Di Mauro 2009). This theory framed cities, including their environments and environmental impacts, as part of global capitalist dynamics in which state-socialist systems, now mostly extinguished, actively participated (Chase-Dunn 1982; Frank 1981; Mies 1986).

As already noted, none of these approaches address environmental processes in themselves as an ecosocial approach strives to do. Therefore, our analysis of urban cultivation includes a specification of the ecological parameters of urban lands. For this, we rely largely on technical works from soil, atmospheric, and biological (ecological) sciences that have been developed specifically for urban contexts. We also build on fledgling critical physical geography work (McClintock 2015) and on the few political ecology exemplars (Pelling 2003) in whose work biophysical processes are investigated instead of being assumed as a given and unchanging substrate.

Ecosocialism

So, then, what is ecosocialism, and what role might urban cultivation have in building it? Ecosocialism is a movement, perspective, and by now even an institutional politics that draws from and attempts to draw together socialist and environmentalist objectives (Pepper 2003). It is socialist or "red" in the sense of identifying capitalist relations as the ultimate, systemic cause of structural inequalities and it is environmentalist or "green" in calling attention to the ecologically destructive character of currently conventional ways of living (Kovel 2014; Löwy 2011). The latter is a general critique mostly of industrialised societies, and the focus is on the outcomes of human impact, in the past as well as the present. As in the case of socialism, environmentalist movements have been diverse from the start, ranging from the self-determination and anti-colonial struggles coinciding with conservation and the more recent movements for environmental justice to the now mainstream colonising, capitalism-friendly, misanthropic/populationist, and/or authoritarian sustainability visions and conservation policies. It is the segment of environmentalism that has a social conscience and is sensitive to social justice issues that overlap with socialism and out of which ecosocialism has developed (Benton 1996; Mellor 1992; O'Connor 1998; Wall 2010). What such environmentalism calls for is the development of a society that is ecologically sustainable and that adopts a cross-generational view of justice, which is what many state-free communities have understood for centuries if not millennia. These forms of environmentalism are anathema to capitalism because they imply a brake on the constant churning out of products for sale.

Socialism, on the other hand and as we see it, is the overcoming of capitalism by establishing the conditions for a society of freely associating producers, sharing resources and democratically administering their affairs. It is to be a society without bosses. The means by which such a society is to be reached necessarily vary according to place, due to widely differing social histories and different contexts relative to global capitalist dynamics. In some places, a combined and coordinated parliamentary and anti-statist road is more apt, as in contexts with centuries to millennia of state-based rule under variable modes of production (or social systems, if one prefers). In other places, where colonialism prevails and peoples continue their struggles for self-determination, priority is given to fighting state institutions and replacing them rapidly with institutions from below. Needless to say, as the outcomes of past revolutions attest (e.g. Ayiti/Haïti/San Domingo in 1804, the Paris Commune in 1871, the October Revolution in 1917, the Spanish Revolution in 1936), capitalism will not be overcome without the worldwide coordination of solidarity actions, mutual aid, globally reaching self-defence, and other like efforts. Ultimately, there can be no socialism in one country (because capitalism is intrinsically and endlessly expansionistic) and no authoritarian redistributive or statist socialism as the final goal, since such social institutions directly contradict the primary aims of socialism in its widest sense. In this understanding, the historical differences among socialism, communism, and anarchism are largely anachronistic, vestiges of bitter political rivalries that have ended up undermining (and still do undermine) attempts at overcoming capitalist relations.

All forms of socialism paid (and some obdurately still do pay) very little attention to the multifarious character of oppression, reducing social struggles to abstract, largely binary class conflicts. Hence, gender, racialisation, and heteronormativity have been treated as secondary issues, if they have not been entirely omitted. The relationship between various kinds of socialism and colonisation is also spotty at best (anti-authoritarian communists and anarchists, on this score, have tended to be much more sensitive). At worst, some socialists have even been cheerleaders for or active participants in colonialism or imperialism.

Until the last few decades, all forms of socialism have had an equally problematic record on ecological relations, that is, to relations between us and other forms of life and to the rest of the planet, as well as to relations among other beings and between them and their physical environments. The consequences in state-socialist countries have been opportunistically well-publicised by anti-socialist scholars and in the capitalist press and need not be reiterated here. For the most part, they are self-serving exaggerations, punctuated by outright fabrications and convenient omissions about the far more destructive nature of liberal democracies and other capitalist political systems (Engel-Di Mauro 2017). Moreover, those accounts of socialist states' environmental records are traversed by illogical comparisons with the wealthiest imperialist powers (Weiner 2017). Unlike socialist states, such countries, mainly liberal democracies, have imposed and still do impose environmental destruction on the rest of the world to feed their own economies while giving themselves a green patina.

There is, in any case, a salient state-socialist exception named Cuba, the only example so far of a state-based system where standards of living have been raised in ecologically sustainable ways (Cabello et al. 2012; Moran et al. 2008). Within anarchism, there are also notable historical cases worth following. There is, among other examples, the work of Reclus, whose ideas about people being active and integral parts of landscapes were a precursor to the notion of bioregionalism. Both Engels and Marx were attentive to and decried the environmental destruction intrinsic to capitalism. That sensitivity—present among prominent segments of the Bolshevik leadership of USSR in the 1920s and then largely suppressed by the late 1930s—is gradually being recuperated among many Marxists (Gare 1993). But it must not be forgotten that Marx and Engels never focused on developing ecological principles and certainly, in practice, never placed environmental issues at the forefront of their political activities. The environmental sensitivities of early socialists, including William Morris (1834–1896), did not form a predominant outlook in socialist movements of any stripe until the 1970s, following the emergence of environmentalism.

In much of the rest of the world or in state-free communities that came under socialist states, ecological sensibility was (and in many parts of the world still is) constitutive of deeply held belief systems, livelihoods, and everyday practices; therefore, environmental harm, such as deforestation, immediately inflicts damage to a community, thereby provoking protest and counter-action. Often, this takes the form of decolonisation struggle, but decolonisation is not coterminous with environmentalism, as environmentally woeful results of national liberation struggles demonstrate (Guha 1989).

Conversely, environmental movements in ecologically indebted countries still tend to be remiss on social justice and especially on decolonisation. It has taken much effort from communities of colour, feminists, and Indigenous communities, among others, to awaken the unsensitised to the fact that environmentalism deprived of social justice only leads to more environmental destruction. However, even barring mainstream obsessions with population growth as if it were inevitable (Malthus 1798) or with technological solutions, a combined social and environmental justice activism rarely looks beyond single places or countries, or realises the global interconnections, the specifically capitalist linkages that ultimately underlie problems like local air pollution. Still, in other situations, where livelihood and ecosystem functions largely overlap, there is no guarantee that environmental struggles include egalitarianism on their agenda.

These are among the salient reasons for the necessary Indigenous peoples' and materialist ecofeminist underpinnings of ecosocialism, where the focus is on class as emergent within multiple forms of oppression, especially racism and heteronormativity (Salleh 2009; Turner and Brownhill 2006), and the oppression of other beings (Adams and Gruen 2014; Pellow 2014). These are understood to form part of an overarching globally intrusive capitalist mode of production that wrecks most peoples' and other beings' lives and places many ecosystems in peril (or so radically alters them as to make them barely liveable). As Kovel (2006) put it, without materialist ecofeminist perspectives, there can be no ecosocialism, especially as patriarchal inequalities have deep social and historical roots (Salleh 1997). Similarly, without supporting Indigenous peoples' revitalisation and decolonisation struggles and learning from their environmental and social practices, there can be little hope for ecosocialism. This is because Indigenous communities, at least those that are non-capitalist, state-free, and egalitarian, tend to be at the forefront of anti-capitalist struggles and tend to preserve or develop ways of life that are ecologically constructive and sustainable. This is not to suggest uncritical support or adoption of their norms and practices. As the late Harold Barclay (1990) understood, oppressive sets of social relations also exist among some state-free peoples, and different modes of oppression may also have developed recently because of upheavals related to colonial encroachment or other invasive processes from capitalist societies (see also Clastres 1974). Developing an ecosocialist society should not therefore be confused with the development of Indigenous peoples' lifeways, whose modes of production are anyway diverse and not necessarily mutually or internally harmonious (this is yet another major challenge ecosocialists must face, but this aspect cannot be taken up in this book). Building ecosocialist movements means that the multiple forms of social and ecological alienation, as insightfully identified by Karl Marx (1844) for capitalist societies (e.g. the process of distanciation described earlier), must be overcome (hence the crucial importance of Indigenous peoples' perspectives and practices). Ecological and social concerns need to be emphatically fused and cross-contextual strategies of mutual support be delineated and developed, so that ecologically sustainable egalitarian communities can germinate, flourish, and diffuse worldwide.

Ecosocial cultivation: developing urban food production to build ecosocialism

Urban food production can play a key role in the construction and development of ecosocialist movements and practices if steered towards ecosocial modes of cultivation. By this we mean urban food-producing communities becoming mindful of and acting upon both ecological and social dynamics simultaneously. It implies multiple kinds of knowledge that are not possible for single individuals to have, and this therefore necessitates drawing in or cultivating people with expertise in both environmental and social matters, who can act as interlocutors and translators among various community members with differing backgrounds and levels of comprehension about the ecological and social aspects of cities and of growing food. This, in turn, implies strong bonds and mutual trust among activists and with the communities in which they are involved. One major difficulty is that knowledge produced and transmitted about ecological and social processes may or may not be mutually intelligible because of the differences in what one pays attention to most and because of the wide range of backgrounds and experiences people tend to have.

More importantly, ecosystems are much larger operating processes than cities, involving many different species and physical processes, so the scale of analysis is necessarily different. What makes for a functioning ecosystem does not necessarily translate into social justice, and the converse can also be true. This is to touch only on the more general complexity to be confronted. In urban environments, ecosystems are radically transformed and often made into toxic soups where some species may thrive and many others may perish or be otherwise harmed, including us. An ecological grasp of cities is fundamental to catching what is harmful and what functions in healthy ways, that is, in ways that enable us as well as other beings to exist in mutually beneficial ways. But it means acquiring and developing technical knowledge and having access to rather resource-intensive equipment, as well as enabling those without knowledge or equipment to gain a sufficient understanding of an urban ecosystem to participate meaningfully in decision-making processes.

Similarly, not everyone has the same degree of attentiveness to relations of power, knowledge of methods that enable egalitarian practices, and understanding of social theoretical issues pertaining to producing and consuming food as well as to urban processes more broadly. Sensitisation to power relations is an essential minimum for those with greater technical proclivities and knowledge. Overcoming relations of power in places where egalitarianism prevails is obviously irrelevant, but it is an arduous challenge in entrenched capitalist settings. The cultivation of resource sharing provides a way forward by establishing and diffusing communal arrangements; this is of fundamental importance in achieving ecosocialism.

One aspect of urban cultivation that leads in that direction is the formation of urban community gardens. These, if actual commons are put in place rather than allotments or similar endeavours, can form a backbone of new, post-capitalist relations (Federici 2012). However, for such commons to function without feeding into capitalist relations at the city scale and without harming communities and ecosystems elsewhere, urban community gardens gain from ecosocialist or like principles and objectives, whether explicit or not. In other words, urban community gardens can become coordinated and confederated, as well as united with struggles for the commons in the countryside. (No Pilgrim or Dakota oil pipelines in North America are possible without the complicity of cities, for example.) They need to focus more directly on issues other than food production, including environmental praxis (i.e. combining theory and practice) and community leadership development (leadership as setting a positive example, not as ruling over people). Furthermore, the management of existing legacies of pollution will be much more just when resources to reduce or (ideally) prevent exposure to toxins are communally controlled, and urban planning and healthcare provisions are communally supported and arranged. Future toxicity is less likely, too, when people feed themselves, even if only in part, by means of urban cultivation.

One objective could then be to turn metropolises and their peripheries into confederated communes, starting at the neighbourhood level, much as the Democratic Confederalism of Rojava (largely in Northern Syria) or in the manner of the Haudanosaunee Confederacy (also known as "Iroquois"). The development of confederated communes, within and beyond cities, will help overcome at the very least those forms of alienation pertaining to a current lack of ecological understanding and sensibility in capitalist societies, as well as the social alienation intensified by capitalist atomisation. Urban cultivation can be a means to achieve these aims, but, in our view, only as part of and consistent with broader ecosocialist aims.

About this book

The following chapters consider the ecological and social aspects of urban food cultivation. They are the fruit of much reading from the works of others who have provided the overviews, insights, and examples from which we draw. We pay special attention to and critically analyse secondary data and reports on some of the more celebrated cases of urban agriculture—Havana (Cuba) and Rosario (Argentina)—as well as some of the ones less publicised—Dar es Salaam (Tanzania), Potchefstroom (near Johannesburg, South Africa), and Tamale (Ghana). Dar es Salaam, Havana, and Tamale feature as part of in-depth ecosocial explorations. Potchefstroom serves as an example of prevalent ideologies that need to be overcome as well as an example of what scientists should not do. Rosario is evaluated with respect to the contributions of urban agroecology relative to environmental and health concerns. Other case studies are the result of our own research and fieldwork in Chongqing, London, New York City, Rome, and San Francisco. Taken together, the case studies considered in this book involve ten cities over four continents as shown on the map. The map is also set as a visual challenge to the dominant ideas about what the world is about, especially with the fixation on national state boundaries and north being necessarily tied to the Arctic.

Our research projects have been methodologically diverse, but they converge on a concern for both the social and environmental justice aspects of urban food production. Some of the studies conducted were more concentrated on the biophysical processes, especially as related to contamination from heavy metals. In New York City and Rome, soil and vegetable sampling and laboratory tests were combined with informal discussions and workshops with urban gardeners, including

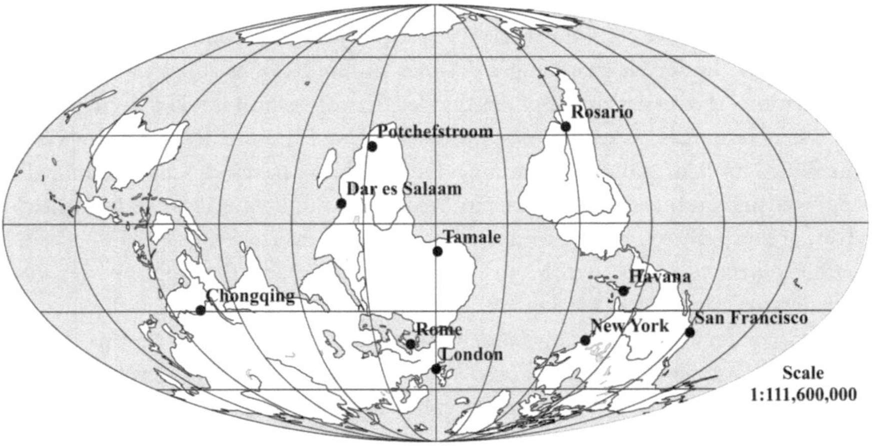

FIGURE 1.1 Case study locations.

Chongqing, London, New York City, Rome, and San Francisco are where the authors carried out fieldwork; Havana, Dar es Salaam, and Tamale provide cases analysed in depth through secondary sources; and Potchefstroom and Rosario are discussed as brief critical commentary on specific topics.

activists. In New York City, the research project, lasting between January and September 2013, saw the participation of ten community gardens in Central and Lower East Side Manhattan and three in Brooklyn (Engel-Di Mauro 2020). Two of the gardens are of historical significance, existing since the 1970s. They were initiated by squatting land and were thereby part of a protracted struggle that led to official recognition and even some gardening support by local government (compost, water access, etc.). The project involved a total of 27 community gardens, inclusive of localities in Albany, Syracuse, and Troy (in New York State), not included in this book. In Rome, research was done from May to August 2014 at gardens established since the early 2000s by activists in the squatters' movement. The focus was on discerning airborne sources of arsenic and lead contamination from among other possible ones (Engel-Di Mauro 2018). For Chongqing, as reported in Rock et al. (2017), there were 37 anonymised interviews carried out in July 2015 with the aid of undergraduates and translators. The activity was supplemented by observations and field analyses of soils and vegetables in 30 gardens (as much as could be achieved without instrumentation and lab work). For New York City and Rome, we cannot report direct quotes from gardeners because no formal interviews were done and therefore no academic Internal Review Board permission was sought (or necessary) for those research projects. With respect to the Chongqing case study, gardeners did not wish to be recorded and reliance on interpreters meant that only essential information from semi-structured interview questions could be transcribed on paper during the interview process.

A second research project involving New York City added other field sites. It began with a site in New York City (Martin 2011) and then one added in London (Martin et al. 2014) and a third one in San Francisco (Martin, Clift, and Christie 2016). Site visits conducted in 2011–2015 consisted of informal in-person interviews with gardeners as well as observation and data collection as to how gardens functioned. The sites were a community garden in the core of New York City, a community farm in the suburbs of London, and an agricultural park in the exurbs of San Francisco. The field analyses of food production, supported by generic estimates, demonstrated very low outputs relative to the populations of the sites' catchment areas. However, all three supported extensive local programmes which contributed to social sustainability, especially in ecological education. The first study involving just the community garden in New York City (Martin 2011) also focused on its historical and demographic history from its 1970s origin forward and included extensive interviews with several of its founders as well as tabulation of demographic data from the US Census Bureau and the Borough of Manhattan.

The contents of the chapters in this book are therefore supported and inspired by conversations with urban gardeners, as well as with left-wing activists who see in urban gardening one basis for overcoming capitalism. They are also, in part, reflective of empirical measures of contamination processes of soils and vegetables as well as of food outputs within community growing plots. This is how the ecological and social have, to some extent, been intertwined throughout this book.

Chapter 2 is an overview of the historical relationship between food production and cities. It turns out that cities and agriculture have not necessarily arisen together. Sometimes, early agriculture had no ostensible relationship with the rise of cities. This is especially evident with the evolution of herding and husbandry. The converse has also been true. There is nothing inevitable or necessarily direct, in other words, in the relationship between city and food procurement strategy.

The recent rise in popularity of urban food production is largely a phenomenon of global North cities. Urban agriculture has hardly been interrupted in most cities in the rest of the world wherever farming co-developed with the urban. Chapter 3 therefore consists of a critique of the largely unscrutinised urban agriculture bandwagon which currently defines a presumed reconciliation between cities and food growing. Behind this revival is the interaction of social structural changes in cities of the global North: the environmental movement, de-industrialisation, urban regeneration through gentrification, in-migration, and foodist/localist cultural trends.

Chapter 4 is dedicated to evaluating how and to what degree urban food cultivation relates to the protection of public health. Wider biophysical factors are incorporated, and soil contamination processes form the primary entry point of analysis. As part of breaking with reductionist understandings of biophysical processes, their dynamics are discussed here in terms of how they potentially affect political struggles within and beyond cities. For example, there are under-appreciated ecosocial repercussions of importing large amounts of soil and sediment, or from manufacturing soils, to grow food in cities. They may promote plant growth, but they may also be environmentally unsustainable as well as unhealthy.

Chapter 5 focuses on the present and potential scale of food growing in cities. An argument is made that it is and will remain scant, consisting of nibbles of food—in absolute volume and relative to the size of local populations. However, a scant output by no means represents a general failure of urban food growing. Even through meagre, food outputs can make meaningful contributions to social sustainability, and to environmental and food justice via their development of local social capital. Urban cultivation can be a means for promoting individual and group learning, public health, and community leadership, which is the subject of Chapter 6. Social capital rather than food is urban cultivation's potential contribution to ecosocialism. In Chapters 5 and 6, site research studies of urban cultivation in three cities of two countries in the global North—London, New York City, and San Francisco—are the direct empirical bases for our analysis of food outputs. Secondary sources are also frequently cited.

In Chapter 7, five cities in five countries are examined: Tamale (Ghana), Dar es Salaam (Tanzania), Rome (Italy), Chongqing (China), and Havana (Cuba). The comparisons aim to assess the degree to which producing food in cities can meet people's needs, not just nutritionally, but also—as has been stressed by many of its advocates—physiologically and socially. The main issue addressed is whether urban food cultivation can serve to promote more egalitarian food distribution, and if so in what ways. In that chapter, the processes described in Chapters 1 through 6 are assembled and integrated. The result undergirds our explication of alternative, ecosocialist food systems and food security measures which can also serve the regeneration of the food and city bond.

In the concluding chapter, we outline an argument for what needs to be done in order to implement an ecosocialist agenda based on our political ecological analysis of urban cultivation and its relation to food systems and food security. A food system approach can integrate rural and peri-urban agriculture with urban cultivation, mitigating the food–city separation. Food security and sovereignty would address the pressing problems of food inequalities (feast or famine) and the threats presented to both food and city sustainability by global climate change. We hope, with this book, to present a path forward to an ecologically sustainable and socially just relationship between cities and food, between urban cultivation and rural agriculture.

References

Adams, C.J., and L. Gruen. 2014. Groundwork. In, *Ecofeminism. Feminist Intersections with Other Animals and the Earth*, edited by C.J. Adams, and L. Gruen, 7–36. New York: Bloomsbury.

Amin, S. 1974. *Accumulation on a World Scale*. New York: Monthly Review Press.

Barclay, H. 1990. *People without Government. An Anthropology of Anarchy*. London: Kahn & Averill.

Beach, Michelle. 2013. *Urban Agriculture Increases Food Security for Poor People in Africa*. Washington, DC: Population Reference Bureau. www.prb.org/Articles/2013/urban-agriculture-poor-africa.asp (Accessed 30 January 2014).

Bell, S., R. Fox-Kämper, N. Keshavarz, M. Benson, S. Caputo, S. Noori, and A. Voigt, eds. 2016. *Urban Allotment Gardens in Europe*. London: Routledge and Earthscan.

Benton, T. 1996. *The Greening of Marxism*. New York: Guilford.

Bernstein, J.H. 2015. Transdisciplinarity: "A Review of Its Origins, Development, and Current Issues." *Journal of Research Practice* 11: Article R1. http://jrp.icaap.org/index.php/jrp/ article/view/510/412 (Accessed 10 August 2020).

Blaikie, P. 1985. *The Political Economy of Soil Erosion in Developing Countries*. London: Routledge.

Cabello, J.J., D. Garcia, A. Sagastume, R. Priego, L. Hens, and C. Vandecasteele. 2012. "An Approach to Sustainable Development: The Case of Cuba." *Environment, Development and Sustainability* 14: 573–91.

Cai, Y., and Z. Zhang. 2000. Shanghai: Trends towards Specialised and Capital-Intensive Urban Agriculture. In, *Growing Cities Growing Food*, edited by N. Bakker, M. Dubbeling, U. Sobel-Koschella, and H. de Zeeuw, 467–77. Feldafing: DSE-ZEL.

Chase-Dunn, C.K., ed. 1982. *Socialist States in the World System*. London: Sage.

Clastres, P. 1974. *La Société contre l'État. Recherches d'Anthropologie Politique* [Society against the State: Political Anthropology Studies]. Paris: Les Éditions Minuit.

Dawson, J.C., and A. Morales, eds. 2016. *Cities of Farmers. Urban Agricultural Practices and Processes*. Iowa City: University of Iowa Press.

Despommier, D. 2018. Vertical Farming Using Hydroponics and Aeroponics. In, *Urban Soils*, edited by R. Lal, and B.A. Stewart, 313–28. Boca Raton, FL: CRC Press.

Eizenberg, E. 2013. *From the Ground Up. Community Gardens in New York City and the Politics of Spatial Transformation*. Burlington VT: Ashgate.

Engel-Di Mauro, S. 2009. Seeing the Local in the Global: Political Ecologies, World-Systems, and the Question of Scale. *Geoforum* 40: 116–25.

Engel-Di Mauro, S. 2014. *Ecology, Soils, and the Left: An Ecosocial Approach*. New York: Palgrave McMillan.

Engel-Di Mauro, S. 2017. The Enduring Relevance of State-Socialism. *Capitalism Nature Socialism* 27: 1–15.

Engel-Di Mauro, S. 2018. An Exploratory Study of Potential as and Pb Contamination by Atmospheric Deposition in Two Urban Vegetable Gardens in Rome, Italy. *Journal of Soils and Sediments* 18: 426–30.

Engel-Di Mauro, S. 2020. Urban Vegetable Garden Soils and Lay Public Education on Soil Heavy Metal Exposure Mitigation. In, *Green Technologies and Infrastructure to Enhance Urban Ecosystem Services. Proceedings of the Smart and Sustainable Cities Conference 2018*, edited by V. Vasenev, E. Dovletyarova, Zh. Cheng, R. Valentini, and C. Calfapietra, 221–6. Cham: Springer.

FAO. 2004. *Agricultural Land Use. FAOSTAT Data*. Rome: Food and Agriculture Organization, United Nations. http://faostat.fao.org (Accessed 29 January 2005).

FAO. 2016. Global and Regional Food Consumption Patterns and Trends: Diet, Nutrition and the Prevention of Chronic Diseases. Rome: Food and Agriculture Organization, United Nations. www.fao.org/docrep/005/ac911e (Accessed 25 March 2016).

FAO. 2018. Urban Agriculture. Rome: Food and Agriculture Organization, United Nations. www.fao.org/urban-agriculture/en/ (Accessed 20 January 2018).

Federici, S. 2012. *Revolution at Point Zero: Housework, Reproduction, and Feminist Struggle*. Oakland: PM Press.

Frank, A.G. 1966. *The Development of Underdevelopment*. New York: Monthly Review Press.

Frank, A.G. 1981. *Crisis: In the Third World*. New York: Holmes & Meier Publishers.

Gare, A. 1993. Soviet Environmentalism: The Path Not Taken. *Capitalism Nature Socialism* 4: 69–88.

Gorgolewski, M., J. Kommisar, and J. Nasr. 2011. *Carrot City: Creating Places for Urban Agriculture*. Singapore: The Monacelli Press.

Guha, R. 1989. *The Unquiet Woods. Ecological Change and Peasant Resistance in the Himalaya*. Berkeley: University of California Press.

Harvey, F., A. Wasley, M. Davies, and D. Child. 2017. Rise of Mega Farms: How the US Model of Intensive Farming Is Invading the World. *The Guardian*, 18 July. www.theguardian/environment/2017/jul/18/rise-of-mega- (Accessed 26 January 2018).

Heynen, N., M. Kaika, and E. Swyngedouw, eds. 2006. *In the Nature of Cities. Urban Political Ecology and the Politics of Urban Metabolism*. London: Routledge.

Kovel, J. 2006. The Ecofeminist Ground of Ecosocialism. *Capitalism Nature Socialism* 16: 1–8.

Kovel, J. 2014. The Future Will Be Ecosocialist, Because without Ecosocialism, There Will Be No Future. In, *Imagine Living in a Socialist USA*, edited by F. Goldin, D. Smith, and M.S. Smith, 25–32. New York: Harper Perennial.

Kurfurst, S. 2019. Urban Gardening and Rural-Urban Supply Chains: Reassessing Images of the Urban and the Rural in Northern Vietnam. In, *Food Anxiety in Globalising Vietnam*, edited by J. Ehlert, and N.K. Faltmann, Chapter 7. Singapore: Springer.

Lefebvre, H. 1974. *La Production de l'Espace* [The Production of Space]. Paris: Éditions Anthropos.

Loftus, A. 2012. *Everyday Environmentalism: Creating an Urban Political Ecology*. Minneapolis: University of Minnesota Press.

Löwy, M. 2011. *Écosocialisme: L'Alternative Radicale à la Catastrophe Écologique Capitaliste* [Ecosocialism: The Radical Alternative to Capitalist Ecological Catastrophe]. Paris: Fayard.

Madden, D.J. 2012. City Becoming World: Nancy, Lefebvre, and the Global—Urban Imagination. *Environment and Planning D: Society and Space* 30: 772–87.

Malthus, Thomas. 1798. *An Essay on the Principle of Population*, 8th ed. London: Reeves and Turner. www.econlib.org (Accessed 27 July 2018).

Martellozzo, F., J.-S. Landry, D. Plouffe, V. Seufert, P. Rowhani, and N. Ramankutty. 2014. Urban Agriculture: A Global Analysis of the Space Constraint to Meet Urban Vegetable Demand. *Environmental Research Letters* 9(6). https://iopscience.iop.org/article/10.1088/1748-9326/9/6/064025 (Accessed 10 August 2020).

Martin, G. 2011. *Transforming a Derelict Public Property into a Vibrant Public Space: The Case of Manhattan's West Side Community Garden*. London: Royal Geographical Society Annual Conference.

Martin, G., R. Clift, and I. Christie. 2016. Urban Cultivation and Its Contributions to Sustainability: Nibbles of Food but Oodles of Social Capital. *Sustainability* 8(5): 409.

Martin, G., R. Clift, I. Christie, and Angela Druckman. 2014. *The Sustainability Contributions of Urban Agriculture: Exploring a Community Garden and a Community Farm*. San Francisco: Life Cycle Assessment International Conference.

Marx, K. 1844 [1978]. The Economic and Political Manuscripts of 1844. In, *The Marx-Engels Reader*, edited by R.C. Tucker, 66–135, translated by M. Milligan, 2nd ed. New York: W.W. Norton.

McClintock, N. 2014. Radical, Reformist, and Garden-Variety Neoliberal: Coming to Terms with Urban Agriculture's Contradictions. *Local Environment* 19: 147–71.

McClintock, N. 2015. A Critical Physical Geography of Urban Soil Contamination. *Geoforum* 65: 69–85.

Mellor, M. 1992. *Breaking the Boundaries: Towards a Feminist, Green Socialism*. London: Virago.

Mies, M. 1986. *Patriarchy and Accumulation on a World Scale*. London: Zed Books.

Mies, M., and V. Bennholdt-Thomsen. 1999. *The Subsistence Perspective: Beyond the Globalized Economy*. London: Zed Books.

Mok, H.-F., V.G. Williamson, J.R. Grove, K. Burry, S.F. Barker, and A.J. Hamilton. 2014. Strawberry Fields Forever? Urban Agriculture in Developed Countries: A Review. *Agronomy for Sustainable Development* 34: 21–43.

Moran, D.D., M. Wackernagel, J.A. Kitzes, S.H. Goldfinger, and A. Boutaud. 2008. Measuring Sustainable Development—Nation by Nation. *Ecological Economics* 64: 470–4.

O'Connor, J. 1998. *Natural Causes: Essays in Ecological Marxism*. New York: Guilford.

Orsini, F., R. Kahane, R. Nono-Womdin, and G. Giaquinto. 2013. Urban Agriculture in the Developing World. *Agronomy for Sustainable Development* 33: 695–720.

Pelling, M. 2003. Toward a Political Ecology of Urban Environmental Risk. The Case of Guyana. In, *Political Ecology. An Integrative Approach to Geography and Environment-Development Studies*, edited by K. Zimmerer, and T. Bassett, 73–93. New York: Routledge.

Pellow, D.N. 2014. *Total Liberation: The Power and Promise of Animal Rights and the Radical Earth Movement*. Minneapolis: University of Minnesota Press.

Peng, J., Z. Liu, Y. Liu, X. Hu, and A. Wang. 2015. Multifunctionality Assessment of Urban Agriculture in Beijing City, China. *Science of the Total Environment* 537: 343–51.

Peng, Y., and L. Hu. 2015. Preliminary Study of Effects of the Model of Community-Supported Agriculture (CSA) on Urban-Rural Income Gap in China. *Agricultural Science and Technology* 16: 404–6.

Pepper, D. 2003. *Eco-Socialism: From Deep Ecology to Social Justice*, 2nd ed. London: Routledge.
Pollan, M. 2016. Big Food Strikes Back. *The New York Times Magazine*, October 9: 40–50, 81–3.
Ponting, C. 2007. *A New Green History of the World: The Environment and the Collapse of Great Civilizations*. London: Penguin Books.
Pulido, L. 2000. Rethinking Environmental Racism: White Privilege and Urban Development in Southern California. *Annals of the Association of American Geographers* 90: 12–40.
Purcell, M., and S.K. Tyman. 2015. Cultivating Food as a Right to the City. *Local Environment* 20(10): 1132–47.
Rehm, C., J. Penalvo, A. Afshin, and D. Mozaffarian. 2016. Dietary Intake among US Adults, 1999–2012. *Journal of the American Medical Association* 315: 2542–53.
Reynolds, K., and N. Cohen, eds. 2016. *Beyond the Kale: Urban Agriculture and Social Justice Activism in New York City*. Athens: University of Georgia Press.
Rock, M., S. Engel-Di Mauro, S. Chen, M. Iachetta, A. Mabey, K. McGill, and J. Zhao. 2017. Food Production in Chongqing, China: Opportunities and Challenges. *Middle States Geographer* 49: 55–62.
Rodney, W. 1974. *How Europe Underdeveloped Africa*. Washington, DC: Howard University Press.
Rose, J. 2016. *The Well-Tempered City*. New York: Harper Collins.
Salleh, A. 1997. *Ecofeminism as Politics. Nature, Marx, and the Postmodern*. London: Zed Books.
Salleh, A., ed. 2009. *Eco-Sufficiency and Global Justice: Women Write Political Ecology*. London: Pluto Press.
Southgate, D., D. Graham, and L. Tweeten. 2007. *The World Food Economy*. Oxford: Blackwell.
Thebo, A.L., P. Dreschel, and E.F. Lambin. 2014. Global Assessment of Urban and Peri-Urban Agriculture: Irrigation and Rainfed Croplands. *Environmental Research Letter* 9. http://iopscience.iop.org/1748-9326/9/11/114002/article (Accessed 10 August 2020).
Tornaghi, C., and C. Certomà, eds. 2019. *Urban Gardening as Politics*. London: Routledge.
Tornaghi, C., and M. Dehaene. 2020. The Prefigurative Power of Urban Political Agroecology: Rethinking the Urbanisms of Agroecological Transitions for Food System Transformation. *Agroecology and Sustainable Food Systems* 44(5): 594–610.
Turner, T., and L. Brownhill. 2006. Ecofeminism as Gendered, Ethnicized Class Struggle: A Rejoinder to Stuart Rosewarne. *Capitalism Nature Socialism* 17: 87–95.
UN. 2014. *World Urbanization Prospects*. New York: Department of Economic and Social Affairs, Population Division, United Nations. https://esa.un.org/unpd/wup/Publications/ Files/WUP2014-Highlights.pdf (Accessed 30 January 2018).
UN. 2016. *Livestock Production*. New York: Department of Economic and Social Affairs, United Nations. www.fao.org/docrep/005/y4252e/y4252e07.htm (Accessed 28 January 2018).
USDA. 2017. *Animal Feeding Operations (AFO) and Concentrated Animal Feeding Operations (CAFO)*. Washington, DC: Natural Resources Conservation Service, US Department of Agriculture. www.nrcs.usda.gov/wps/portal/nrcs/main/national/ plantsanimals/livestock/afo/ (Accessed 28 January 2018).
Wall, D. 2010. *The Rise of the Green Left*. London: Pluto Press.
Weiner, D.R. 2017. Communism and Environment. In, *The Cambridge History of Communism*, edited by J. Fürst, S. Pons, and M. Selden, 529–55. Cambridge: Cambridge University Press.
WFP. 2017. *Counting the Beans: The True Cost of a Plate of Food around the World*. Rome: World Food Programme, United Nations.

WinklerPrins, A.M.G.A. 2017. Defining and Theorizing Global Urban Agriculture. In, *Global Urban Agriculture: Convergence of Theory and Practice Between North and South*, edited by A.M.G.A. WinklerPrins, 1–11. Boston, MA: CABI.

Zezza, A., and L. Tasciotti. 2010. Urban Agriculture, Poverty, and Food Security: Empirical Evidence from a Sample of Developing Countries. *Food Policy* 35: 255–73.

2

CITIES AND FOOD PRODUCTION

Cities and food production are increasingly coming together as if they were once long-separated twins reunited at last, as if to rekindle an ancient bond. It may then be surprising to learn that historically the development of agriculture and animal husbandry did not necessarily coincide with the earliest forms of permanent human settlement (Kuijt and Goodale 2009; van der Veen 2005). Eventually, though, farming, animal husbandry, and cities came to be closely associated almost everywhere. This bond has endured in many cities, a reflection of the essential link between the places where food is produced and where it is consumed. Yet it has also shifted in spatial configuration and extent of contiguity. That is to say, there have been major changes within cities in terms of where people produce food and where they live and do other kinds of work. In some parts of the world, especially those where capitalist relations and industrialisation took root, this connection has become increasingly removed physically and socially. This is now even more so as cities have expanded to engulf ever-larger areas and as an increasing majority of people have become urbanised (often by force).

The once seamless commingling of everyday life with other animals and with collecting and making food has been largely overtaken by an often sharp division between town and country. The distancing of animal husbandry from cropland is a corollary to this separation, producing mountains of polluting excrement in some places and exhausted (nutrient-depleted) soils in others. In most regions, landscapes have been churned up to produce more or less contiguous metropolitan areas and then megacities of over 10 million inhabitants. Where but a short stroll could allow one to be immersed in cultivated land or herds, there now often takes hours of sometimes traffic-ridden driving to reach the nearest farm. Both human settlements and farms have required more and more land over time, while many other animals have instead been more and more confined to tinier areas or cages, to satisfy primarily the bottomless pit of capital accumulation (generating profits for

DOI: 10.4324/9781003131281-2

profits' sake). The accumulation is now topped by multi-billionaires who are chasing yet more profits. The result has been the diffusion of ecological destruction, coupled with social estrangement between people, people and their food, people and other species, along with huge nutritional disparities.

None of this is to regret the historical passing of horrific kinds of societies in some parts of the world (e.g. those based on slavery), where despotic relations embittered most people's already tough existence punctuated by plagues, famines, and other such disasters. However, it is appropriate to mourn the disappearance of Indigenous societies that were ecologically sustainable and politically egalitarian, and to extoll those persisting to this day against enormous odds. Existing ecologically sustainable egalitarian societies may not be amenable to accommodating the realities of globalisation, its megacities, sprawling monocultures, and animal-crippling husbandry. Hopefully, however, those once-thriving and now surviving can help to preserve human diversity as well as to point towards paths out of capitalism's exploitation of most people and the rest of nature. They can, along with people struggling within capitalist societies, help to check and eventually to overcome the historically expansionistic tendency of capitalist relations, headed by what are essentially ruling minorities ultimately defended by armed thugs (Tilly 1985; Wallerstein 1991, 33–5). Capitalist profiteering is underlain by an unparalleled destruction of ecosystems headlined by toxic fossil fuels and hazardous chemicals that contaminate the essential foundations of human life—air, soil, and water—jeopardising public health and promoting climate change threats.

From town and country to urban and rural

By depriving most people of their sources of subsistence (the primordial result of privatising property), capitalist relations begot the sundering of town and country and with that the estrangement of millions of urban dwellers from the origins of their food. Increasing separation does not by any means reduce the linkages between food and city, which are as inextricable as ever. Agricultural production in the country and consumption, mostly in cities, intersect across multiple facets of daily life—personal and social, dietary and symbolic, cultural and political. To state the obvious, food enables our material existence, so its ecological and physiological bases are crucial to us. Its production, even in the most mechanised and agrochemically capital-intensive versions, requires minimally adequate biophysical supports—soil, sunlight, water, amenable weather, and breathable and plant-friendly air. At the same time, social relations shape what is grown where and how and they ultimately determine who benefits from food production and its distribution, while what is designated as food is mediated by cultural practices.

Technically, the emergence of a sedentism and food-growing bond constituted a momentous change in the ecological evolution of some societies. Over thousands of years, this bond changed radically and diversified as it spread unevenly far and wide and transformed landscapes (Mazoyer and Roudart 2006). Many changes followed that enabled increasing food surpluses and higher-density urbanism, including rice-paddy systems, intercropping, hydraulic systems (like irrigation channels), metal farming tools, mountain-slope terracing, granaries, and so on (Braudel 1982).

A more recent momentous technological change—industrialisation—came only about 250 years ago, as part of the social convulsions and catastrophes that came to be known as capitalism. Following the largely coercive and genocidal imposition of capitalist institutions nearly everywhere, all parts of the world are still engaged with industrialisation. In many places, an industrialising process is still happening, largely under terms set by the hegemonic global North nations that are de-industrialising by relying on cheaper imports from, or displacing manufacturing to, developing nations with low labour costs.

This is one way that cities have become differentiated in their historical trajectories to reach extremely unequal levels of consumption and waste generation and of waste distribution. That is, it can be useful to keep in mind that cities have eventually split according to general wealth levels and degrees of internal inequality (Brambilla, Michelangeli, and Peluso 2015). Parsing cities, according to conventional notions of development (e.g. based on indices like the Human Development Index), is therefore inadequate if not spurious. In such conceptualisation, cities in developing countries are deemed to share the problems of prioritising development over environmental policy, inadequate or absent integration of environmental management and urban planning, and a dearth of spatial databases (Cilliers, Bouwman, and Drewes 2009, 110). The reason this methodology falters is that levels of resource consumption and waste tend to correlate with wealth accumulation, such as investments (local, national, and international), sales, incomes, and the like. Development indices ignore activities external to capitalist logic—e.g. subsistence-oriented production, informal educational systems, communal resource-sharing mechanisms, degree of egalitarianism within households, and redistribution through bartering. Grouping cities according to conventional indices of development is also directly contradicted by substantive divergence among cities within developing countries according to the degree of general economic dependence on raw material exports. Moreover, historical primary sector predominance tends to feature services-oriented urban development and lower levels of material well-being compared to cities where manufacturing represents a larger share of the national economy (Gollin, Jedwab, and Vollrath 2016).

This more recent epochal change in the relationship between cities and food and the vast developmental inequalities between and within cities has a well-known prime mover. It is capitalism as a mode of production, with "its abstracted economic drives, its fundamental priorities in social relations, its criteria of growth and profit and loss" (Williams 1973, 302); or, otherwise put, with its hegemony over how most of us can live—a hegemony attributed to structural abstractions like the "market", "commodities", "debt", and "money" (Adamczak 2017). This mode of production was established at first through changes in the social relations of farming. Its historic hallmark has been uneven development spearheaded by colonial empires. Industrialisation made its initial impacts on agriculture, and its influence on food production and distribution has continued. The coupling of capitalism with industrialism and urbanism led to the emergence of what is referred to as "modern" societies, or, as we will call them here, capitalist societies.

The evolution of the food–city relationship has featured contradiction and variability. Over the past several centuries, as capitalist relations emerged and were imposed globally, the commodification of food into monetary tradeable status has led to highly uneven food consumption, with repercussions ranging from chronic overfeeding in some populations to recurring famines in others. Among the shortcomings of today's global agriculture, the headline is "that 1 billion people are malnourished, more than 1 billion are over-nourished (and overweight) and health services around the world are dealing with rising levels of diet-related ill-health" (Sage 2012, 205). Despite large-scale global food insecurity, the world "already produces enough edible calories to feed" all of humanity (Chappell 2018; cited in Vandermeer et al. 2018, 39). In fact, conventional capitalist farming has failed, even according to mainstream institutional terms, as shown in a recent international scientific report (IAP 2018, 6–7).

While food growing has evolved into an activity largely reserved for what became known as the rural, it is, in most of the world, also a part of urban life. In most cases, cities have never ceased to include food production within their boundaries, even if under highly variable policies, sometimes favourable and sometimes restrictive. Where industrialisation and cites have been expanding, urban food-producing areas have typically been diminished. Sometimes they have been drastically downsized or banished from urban centres, as in China, where fast-paced industrialisation, commodification, and privatisation have led to net ecological deficits within a mere three to four decades. By contrast, in the cities of the much fewer and more ecologically indebted countries of the developed global North, food growing has waxed and waned with economic cycles and wars. In these cities domestic gardens and allotments have been the main urban production sites and both have declined over time. Food growing has only recently been revived by community gardens, stimulated by different factors that include the effects of de-industrialisation (McClintock 2014). The resurgence has led to a change in the status of urban food growing—it is now increasingly referred to as urban agriculture.

But what is urban agriculture and how does it differ from horticulture or gardening in cities? The FAO defines urban agriculture as growing plants and raising animals within and around cities to provide fresh food, generate employment, recycle waste, and strengthen cities' resilience to climate change (FAO 2018). This definition highlights a general expert consensus about the important global role of urban food growing in the establishment of sustainable cities. We are taking this definition as the starting point for critiquing urban agriculture and proposing an ecosocialist alternative. The cornerstones of these alternatives will be social and food justice as well as ecological sustainability.

Agriculture, permanent settlement, and built environments

Nowadays, it is taken for granted that urbanity and agriculture are connected, even if many people might not know where exactly their food comes from. From about 9,500 to about 5,000 years ago (or BP, Before Present), agriculture and settlements developed synergistically within some areas in South and Southwest Asia and Southeast Europe (Manning 2008). In such places, internal contradictions and major social changes, especially towards a predominance of patriarchal oppression (Hughes and Hughes 2005; Lerner 1987), led to a more centralised control of food surplus and its allocation (Harman 2002, 22–6). These social contradictions and changes were likely crucial to the emergence of political systems such as kingdoms and empires and ideological institutions to justify them. Permanent hierarchies and centralised chains of command, notably bureaucracies, were organised to facilitate the ruling over of ever-larger populations and the management of greater food supplies, often achieved by military conquest and impositions of tribute. One of the enduring legacies of these kinds of ancient agriculture–city bonds has been patriarchy and various forms of slavery or coerced and indentured labour. The first, gender-based authoritarianism enabled the population expansion that provided the basis for ever-larger militaries and production levels (not only in agriculture) as well as the intensification of technical divisions of labour. The second legacy, slavery and other forms of coerced labour, established the means to marshal the enormous amount of physical labour required for surplus food production, including physical infrastructures such as irrigation systems and cities with their monuments, roadways, and fancy residences for the ruling strata (Anderson 2014; Standage 2009). It was not until the late 1400s, however, that such authoritarian societies began to expand to such an extent as to eventually engulf the entire world—an expansion based in Western European capitalist colonialism (Amin 1974; Frank 1978; Mies 1986; Wallerstein 2011).

Yet the historical relationship between food growing and cities (as a kind of sedentism) is hardly that straightforward in most of the world (Manzanilla 1997). The timing and location of the earliest cities are as uneven as those characterising plant and animal domestication. The spread of cities and agriculture and their mutual dependence, which, we now take for granted, took thousands of years to take shape. The earliest permanently settled areas seem to date as far back as 11,000 years ago in Southwest Asia, at sites like Jericho (in today's Palestine, in the West Bank of the Dead Sea). However, permanent settlements there arose before plant or animal domestication. Not too far away, but much later (about 9,400 BP), in Çatalhöyük (in today's Turkey), farming was well developed prior to the formation of the city (Fairbairn 2005). It might seem that as agriculture developed further and spread, cities would have followed. This is the likely sequence in the Indus Valley, such as at Mehrgarh (in today's Pakistan), starting at about 8,500 BP. But for thousands of years in what is now Japan, during the Jomon period, permanent settlement was based largely on fishing and gathering (Pearson 2006).

Across most of North America, farming was seldom associated with the extensive, permanent settlements known in other parts of the Americas, such as the Maya and Inka regions. The historical geography of the earliest cities is like the sprouting and disappearance of tiny dots spread over vast tracts of landmasses.

The emergence of farming is just as, if not even more, complex. The division between agriculture and horticulture does not apply so easily in the earliest forms of farming. The differences were, and still can be, more about relative amounts of human labour and other inputs per unit of land area and what makes most sense to do relative to local ecosystems and prevailing social relations (van der Veen 2005). For instance, shifting cultivation allows for more adaptability in tropical forest conditions, but this would not be a very effective set of techniques in arid lands or grasslands. What is considered gardening now is also likely to have been the extent of farming thousands of years ago. What is large scale for one period and one society may be small scale in another period or to another kind of society. What is also to be borne in mind is the multiplicity of origins, the lack of synchrony, and the major differences in the types of cultivars and technologies in the development of agriculture. The earliest evidence of domestication (not yet farming, it must be stressed) is at 10,000 BP in Southwest Asia, with wheat and barley, and Meso-America (present-day Southwest México), with squash and then maize and beans (thousands of years apart). In other areas, namely Western Amazonia, the Central Andean region, China, New Guinea, Eastern North America, and Africa's Sahel, thousands of years later, entirely different kinds of plants were domesticated and different kinds of farming emerged (Bellwood 2005; Price 2009; Smith 1994). An analysis of the first, Neolithic forms of agriculture found a similar emergent foundation across continents and over thousands of years: changing environmental conditions to suit cultivation and, sometimes (depending on the outcomes of social relations of power), enable higher population densities (Kavanagh et al. 2018).

This is not even to start on the development of non-human (often vertebrate) animal domestication, which first involves wolves (becoming dogs) about 13,000 BP, if not earlier. Domestication, in this case, is more related to foraging and hunting than farming, and dogs rarely became meat. As with plant domestication, subsuming other animals under food procurement evolves very gradually, over thousands of years, and with great geographical diversity in timing and species. The relationship between animal husbandry and sedentism is tenuous at best. Many societies developed close relationships with other animal species (not only mammalian or even vertebrate, like bees and silkworms) without their domestication being associated with cities. In fact, in Central Asia and Siberia, for example, people came up with elaborate and ingenious ways of ensuring pasture and water for herds in combination with foraging and hunting (including fishing), spanning hundreds of kilometres of distance and relying on semi-permanent and temporary encampments. The connection between city and husbandry is indirect, with specific forms of animal domestication more related to combined city–agriculture complexes (e.g. cats, fowl, and pigs) and others evolving out of nomadic or semi-sedentary ways of life (e.g. cattle, water buffalo, camels, sheep, goats, and reindeer).

There is therefore no historically necessary bond between the city and agriculture, and much less with animal husbandry. Another kind of major social change was necessary to forge a close association between farming and cities (animal husbandry only partially gets tangled up in this). Arguably, it is not until substantive social stratification (inequality) and the formation of empires that the bond between cities and farming became inescapable. This is corroborated by a historical analysis of 155 Austronesian societies in Southeast Asia, where it was concluded that agricultural intensification and social hierarchy co-evolved (Sheehan et al. 2018). The earliest such development was at about 5,000 BP in Mesopotamia and Kemet (Egypt), and then at about 4,000 BP in the Indus Valley and 3,800 BP in North China. In other places, empires formed much later, as in Meso-America (2,100 BP) and Perù (1,500–1,700 BP). Such polities (often conflated tendentiously with "civilisation") were initially very much circumscribed exceptions, also as city-states, and they continued to be geographically limited exceptions for thousands of years.

Eventually, the mix of states, agriculture, and cities congealed in many regions, but even in this case linkages are highly uneven. Farming existed for thousands of years in many areas were no state ever formed, but was rather imposed, as in the Sahel and Western Amazonia. Even more recently, no such necessary connection existed. The Mongol Empire was based on the centralisation of authority but was not combined, as was the Kemetian, Inkan, and Roman Empires, with large, permanent settlements. The historical tendency may have been, at least for some areas of the world, a mutually reinforcing development of sedentism and cities, farming and (to some extent) animal husbandry, and social inequality and centralised power, but tendencies do not imply any inevitability or preordained order, nor necessarily any durability or stability.

In short, there were multiple, independent formations of states, and they were far from the sort of pervasive, life-intrusive states we know today. Cities, in some cases, led to state formation and in others not, and sedentism and farming were not necessarily related processes. What many current narratives still miss is that it was with the "modern" or national state that sedentism, cities, structural social inequality, political centralisation, farming, and animal husbandry were fused. This came about only a couple of centuries ago with the development of much more centralised and ever-more lethally powerful states than any before them (Harman 2002, 12; Scott 2017; Tilly 1990). Such states, identified as cradles of (liberal) democracy, formed in parts of Europe and expanded on the basis of genocides, slavery, and colonialism (Losurdo 2005).

The new domains of food production and consumption, largely divided between country and town, have their origins in the advent of the predominance of capitalist relations (García-Sempere et al. 2018), with their highly uneven development, rise and spread of industrialisation and urbanisation—all based on an inherent imperialism of the more economically and militarily powerful national states. These historical changes bequeathed us the creation of globally commanding but ecologically indebted countries by means of the appropriation of labour, raw materials, and food commodities (Hornborg 2011; Salleh 2009). As for environmental sustainability, there has also been, since the late twentieth century, a newly recognised aspect of resource appropriation—that is, appropriated from future generations as well. The urban–rural dichotomy today is an actively contributing product of these historical developments, a reflection and reproducer of the intrinsic and violent incoherence of capitalist relations, but also simultaneously and paradoxically one among many seeds useful towards building constructive relations among us, as well as between us and the ecosystems of which we are part.

Those who claim that agriculture was the result of environmental change or that it is the ultimate culprit of present environmental ills share a penchant for inverting historic realities. They cover up the social causes (relations of domination, stratification) behind the emergence of ecologically destructive forms of farming by papering over the huge chasms in the ecological impacts of different kinds of farming. They impute mono-causal power to a multiplicity of often unrelated eco-social effects of diverse forms of farming. Such positions reveal a thoroughly reactionary politics in that attention is diverted away from the gross social injustices on which industrialised, profit-driven agriculture is predicated. It is politics that conforms to the ideological ends of the current ruling classes, which is to dissimulate the capitalist farming roots of many coupled environmental and social disasters.

Estrangement between cities and food production

The agriculture–city bond is now commonly viewed to comprise two grand abstractions—the extensive production of food in rural areas and its large-scale consumption in urban areas. The two realms are socially as well as physically distanciated. Social relations in each sphere are rooted in diverse cultural practices as well as variable landscapes and topographies. The differences have become part of common-sense capitalist ideology that also permeates the sciences. They are the basis of a foundational paradigm in the nineteenth-century modernist emergence of social science—the provincial rural community and the cosmopolitan urban society (Tönnies 1887). While provincial is a designation dating from feudal Europe, cosmopolitan is relatively new, first used as an extension of the term metropolitan, reflecting the expanding urbanism and worldliness of modernism.

The physical and social food–city separation was widened and intensified with the development of industrial production in cities. For example, at the beginnings of the industrial revolution in the mid-1700s, London was relatively compact and proximate to its food sources; it occupied about 40 km^2 in area (ca. 15 mi^2) and had a population of some 700,000 souls. By the end of the revolutionary period of industrialisation, in 1851, London's population had quadrupled, and its area had increased by a multiple of 40. Present-day Greater London, which includes suburbs, is 1570 km^2 (607 mi^2) with a population of 8.2 million. The London region or megacity, reaching to its exurbs, is about 70,000 km^2 (27,000 mi^2) and has a population of 20.3 million (Cox 2012; Emsley, Hitchcock, and Shoemaker 2018; Roumpani and Hudson 2014). Sizeable farms once stood in what are now London's suburbs and exurbs, replaced over time by single-family homes, offices, etc. Today, over one-half of the UK's food comes from abroad, 30 per cent of it from European Union countries (DEFRA 2016). In the US, the transition from traditional craft to modern mass production over the course of the twentieth century incorporated three major technological developments, from mechanisation through chemicalisation, to the contemporary establishment of biotechnology. In the same century, from 1910 to 1997, the number of farms was reduced by two-thirds while the average farm size increased threefold (Lyson 2004, 19–21), while the population grew from being 46 per cent to 78 per cent urban (USBC 2017).

A major contributor to the city and food distanciation process has been the industrialisation of agriculture, which has resulted in the use of food production technologies based on fossil fuels, pesticides, synthetic fertilisers, seed hybridisation, and genetic engineering. These alone, irrespective of all the intervening processes separating consumer from producer, have compelled recourse to ever-larger sums of capital in order to be able to grow food. The massive shift away from traditional farming to modern agriculture was underwritten historically by forced land clearances (Fairlie 2009). This was a central aspect of what Marx (1967) analysed in Capital under the term primitive accumulation. Many of the people evicted from the land found their way to cities to become part of the industrial working classes. The transition from agricultural- to industrial-based societies was the historic lever for a massive transfer of labourers from (rural) farms to (largely urban) factories. The industrial revolution began in fact with a focus on the countryside. Central to its development were the manufacture of steam-powered farm machinery and the production of textiles by women in their farm homes—the "putting-out" system as it was named in England. When textile mill factories were built in cities, they served as magnets for the migration of work-seeking farmers who had been displaced by clearances and the mechanisation of agricultural labour.

As a result, the urban population of England and Wales surged during the industrial transition, growing from 34 per cent of total population in 1801 to 50 per cent in 1851 and to 78 per cent in 1911 (Law 1967). Where England and Wales led, the rest of the world followed. In 1800, only 3 per cent of global population lived in cities; in 1950, 30 per cent and in 2014, 54 per cent; this is projected to increase further, to 70 per cent by 2050 (UN 2014). While earlier humanity transitioned from scattered gathering and hunting to a settled food growing, a second and still current transition is industrialisation coupled with full-on urbanisation, resulting in ever-more estrangement between people and their commodified food sources.

The expansion of cities and decline in their adjacent food growing is accelerating globally as cities grow ever larger. Research on land-cover change for 50 cities around the world showed that their expansions between 1985 and 2010 were strongly negatively correlated with changes in neighbouring croplands, forests, and grasslands (Bagan and Yamagata 2014). These green lands were replaced by suburbs and exurbs, which help explain why settlement growth had a weak negative correlation with changes in wetlands—they are not desirable as building sites in an industrialised capitalist society (palafittes or stilt houses, on the other hand, are perfectly feasible in other societies). The settlements having the strongest correlations with adjacent land changes were a mix from the global North (Amsterdam, New York City, Toronto, and others) and South (Bangkok, Beijing, Rio de Janeiro, and others). Massive urban expansion has led to reduction of city-peripheral, small-scale farming and growth of large-scale rural farming. Globally, today, 98 per cent of farms are small-scale (family), and they occupy 53 per cent of agricultural land, while 2 per cent are large-scale (corporate) farms that occupy the remaining 47 per cent (Graeub et al. 2015; cited in Vandermeer et al. 2018).

To reiterate, as human numbers and their urban habitation have mushroomed, the city–food nexus has become increasingly complex and distanciated. Social relations between food producers and consumers have become ever more abstracted and institutionalised while their reach across different countries, groups, and communities has extended. Many profit-driven processes of mediation now intervene between food grown and food eaten, including processing, warehousing, transporting, and wholesaling. The age-old intimate and proximate relationship between a city and its food has become formalised and globalised. Food growing has become a lost craft for many today, especially those in the global North.

Industrialised agriculture has a ravenous natural resource diet, and it has caused extensive environmental damage, much of it from a reliance on chemicals and fossil fuels. Agriculture uses about 70 per cent of the world's land surface, and increased fertiliser use has degraded its quality in many regions of the world (Foley et al. 2005). It accounts for 30 per cent of global greenhouse gas (GHG) emissions. Pre-production emission contributions come from making fertiliser, pesticides, and herbicides (Meyer and Reguant-Closa 2017). Post-production emissions come from packaging, transportation, and waste. Altogether, it is the direct on-farm and indirect (from deforestation) stages of agriculture that contribute the biggest share of food-related GHGs.

The chemical and biotechnical transformation of once naturally diverse food continues apace. A study of 19 European countries (Monteiro et al. 2018) revealed a high level of ultra-processed food in daily diets. These are composed of machine-produced ingredients and additives created in laboratories by food technologists. The UK leads other European nations with a bit over one-half of its families' diets being composed of ultra-processed food. A common one is cheese made of milk power and additives. At some point, it is likely that research will indicate a link between this kind of food and obesity and poor health (Boseley 2018).

Today, consumers find it increasingly difficult to learn the sources and environmental impacts of food. Princen, Maniates, and Conca (2002) analysed the situation to be a result of a distancing process that displaces feedback loops. Given capitalism's demand for continually increasing economic growth, any efficiency and technical gains in food production are counterbalanced by stimulating more consumption achieved through advertising. The model of production–consumption dualism now resembles a treadmill, and not keeping the pace can result in environmental and health calamities for individuals and societies (Lock and Ikeda 2005; Schnaiberg 1980).

To sum up our critical historical analysis of the food–city relationship, domestication, agriculture, and settlement are today fused. They developed asynchronously, over thousands of years, and very differently in different parts of the world. Eventually, with major social shifts towards patriarchal relations and various forms of forced labour, food procurement and dwelling strategies formed the basis for permanent social stratification and early state formation. Out of these historical changes came a close linkage between agriculture, animal husbandry, and urban life in some regions of the world. Over the span of the last few centuries, an estrangement emerged in the food–city bond with the rise and spread of capitalism, based on conquest, genocides, and commodity trading involving natural resources, foods, fibres, and slaves. The globalisation of capitalism developed from this commercialism (merchant) base through industrialism (manufacture), to present-day financialism (corporate) predominance (Gordon 1984). The estrangement has been the basis for a cultural and social country–city (and production–consumption) dualism which was notably analysed and critiqued by Raymond Williams (1973, 296), who put the matter thus: "we must not limit ourselves to their contrast but go on to see their interrelations and through these the real shape of the underlying crisis." The urban food-growing phenomenon of our time represents an emergent rapprochement in the dichotomous food–city relationship and discourse, and it merits our substantive political and ecological analysis.

References

Adamczak, B. 2017. *Communism for Kids*. Translated by J. Blumenfeld, and S. Lewis. Cambridge, MA: MIT Press.

Amin, S. 1974. *Accumulation on a World Scale*. New York: Monthly Review Press.

Anderson, E.N. 2014. *Everyone Eats: Understanding Food and Culture*, 2nd ed. New York: New York University Press.

Bagan, H., and Y. Yamagata. 2014. Land-Cover Change Analysis in 50 Global Cities by Using a Combination of Landsat Data and Analysis of Grid Cells. *Environmental Research Letters* 9. doi: 10.1088/1748–9326/9/6/064015 (Accessed 30 January 2018).

Bellwood, P. 2005. *First Farmers: The Origins of Agricultural Societies*. Oxford: Blackwell.

Boseley, Sarah. 2018. 'Ultra-Processed' Products Now Half of all UK Family Food Purchases. *The Guardian*, 2 February. www.theguardian.com/science/2018/feb/02/ultra- (Accessed 2 February 2018).

Brambilla, M.R., A. Michelangeli, and E. Peluso. 2015. Cities, Equity and Quality of Life. In, *Quality of Life in Cities: Equity, Sustainable Development and Happiness from a Policy Perspective*, edited by A. Michelangeli, 91–109. London: Routledge.

Braudel, F. 1982 [1979]. *Civiltà Materiale, Economia e Capitalismo. Secoli XV-XVIII, Le Strutture del Quotidiano* [Material Civilisation, Economy, and Capitalism. The XV-XVII Centuries: The Structures of Everyday Life]. Torino: Einaudi.

Chappell, M.J. 2018. *Beginning to End Hunger*. Berkeley: University of California Press.

Cilliers, S., H. Bouwman, and E. Drewes. 2009. Comparative Ecological Research in Developing Countries. In, *Ecology of Cities and Towns: A Comparative Approach*, edited by M.J. McDonnell, A.K. Hahs, and J.H. Breuste, 90–111. Cambridge: Cambridge University Press.

Cox, Wendell. 2012. The Evolving Urban Form: London. *Newgeography*, 23 July. www.newgeography.com/content/002970-the-evolving (Accessed 27 January 2018).

DEFRA. 2016. *Food Statistics in Your Pocket 2017—Global and UK Supply*. London: Department for Environment Food and Rural Affairs. www.gov.uk/government/ publications/food-statistics-pocketbook-2017/food-statistics-in-your-pocket-2017-global-and-uk-supply (Accessed 29 January 2018).

Emsley, C., T. Hitchcock, and R. Shoemaker. 2018. London History—A Population History of London. *Old Bailey Proceedings Online*. www.oldbaileyonline.org/static/Population-history-of-london.jsp (Accessed 27 January 2018).

Fairbairn, A. 2005. A History of Agricultural Production at Neolithic Çatalhöyük East, Turkey. *World Archaeology* 37: 197–210.

Fairlie, S. 2009. A Short History of Enclosure in Britain. *The Land Magazine*. www.thelandmagazine.org.uk/articles/short-history- (Accessed 28 January 2015).

FAO. 2018. Urban Agriculture. Rome: Food and Agriculture Organization, United Nations. www.fao.org/urban-agriculture/en/ (Accessed 20 January 2018).

Foley, J.A., R. Defries, G.P Asner, C. Barford, G. Bonan, S.R Carpenter, F.S. Chapin, M.T. Coe, G.C. Daily, H.K. Gibbs, J.H. Helkowski, T. Holloway, E.A. Howard, C.J. Kucharik, C. Monfreda, J.A. Patz, I. Colin Prentice, N. Ramankutty, and P.K. Snyder. 2005. Global Consequences of Land Use. *Science* 309: 570–4.

Frank, A.G. 1978. *World Accumulation, 1492–1789*. New York: Monthly Review Press.

García-Sempere, A., M. Hidalgo, H. Morales, B.G. Ferguson, A. Nazar-Beutelspacher, and P. Rosset. 2018. Urban Transition Toward Food Sovereignty. *Globalizations* 15: 390–406.

Gollin, D., R. Jedwab, and D. Vollrath. 2016. Urbanization with and Without Industrialization. *Journal of Economic Growth* 21: 35–70.

Gordon, David M. 1984. Capitalist Development and the History of American Cities. In, *Marxism and the Metropolis: New Perspectives in Urban Political Economy*, edited by W. Tabb, and L. Sawers, 21–53. New York: Oxford University Press.

Graeub, B., M. Chappell, H. Wittman, S. Ledermann, R. Kerr, and B. Gemmill-Herren. 2015. The State of Family Farms in the World. *World Development* 87: 1–15.

Harman, C. 2002. *A People's History of the World*. London: Bookmarks.

Hornborg, A. 2011. *Global Ecology and Unequal Exchange: Fetishism in a Zero-Sum World*. London: Routledge.

Hughes, S.S., and B. Hughes. 2005. Women in Ancient Civilizations. In, *Women's History in Global Perspective*, vol. 2, edited by B.G. Smith, 9–46. Urbana: University of Illinois Press.

IAP. 2018. Opportunities for Future Research and Innovation on Food and Nutrition Security and Agriculture. In, *The InterAcademy Partnership's Global Perspective*. Washington, DC: InterAcademy Partnership.

Kavanagh, P.H., B. Vilela, H.J. Haynie, T. Tuff, M. Lima-Ribeiro, R.D. Gray, C.A. Botero, and M.C. Gavin. 2018. Hindcasting Global Population Densities Reveals Forces Enabling the Origin of Agriculture. *Nature Human Behaviour* 2: 478–84.

Kuijt, I., and N. Goodale. 2009. Daily Practice and the Organization of Space at the Dawn of Agriculture: A Case Study from the Near East. *American Antiquity* 74: 403–22.

Law, C.M. 1967. The Growth of Urban Population in England and Wales, 1801–1911. *Transactions of the Institute of British Geographers* 41: 125–43.

Lerner, G. 1987. *The Creation of Patriarchy*. Oxford: Oxford University Press.

Lock, I.C., and S. Ikeda. 2005. Clothes Encounters: Consumption, Culture, Society, and Economy. In, *Consuming Sustainability: Critical Social Analyses of Ecological Change*, edited by D. Davidson, and K. Hatt, 20–46. Halifax NS: Fernwood Publishing.

Losurdo, D. 2005. *Controstoria del Liberalismo* [A Counter-History of Liberalism]. Roma: Laterza.

Lyson, T.A. 2004. *Civic Agriculture: Reconnecting Farm, Food, and Community*. Medford MA: Tufts University Press.

Manning, S. 2008. Year-by-Year World Population Estimates: 10,000 B.C. to 2007 A.D. *Historian on the Warpath*, 12 January. www.scottmanning.com/content/year-by-year-world-population-estimates/ (Accessed 29 January 2018).

Manzanilla, L., ed. 1997. *Emergence and Change in Early Urban Societies. Fundamental Issues in Archaeology*. Boston, MA: Springer.

Marx, K. 1967 [1867]. Capital: A Critical Analysis of Capitalist Production. In, *Primitive Accumulation*. Vol. 1, Part VIII, edited by Frederick Engels. New York: International Publishers.

Mazoyer, M., and L. Roudart. 2006. *A History of World Agriculture from the Neolithic Age to the Current Crisis*. Translated by J.H. Membrez. New York: Monthly Review Press.

McClintock, N. 2014. Radical, Reformist, and Garden-Variety Neoliberal: Coming to Terms with Urban Agriculture's Contradictions. *Local Environment* 19: 147–71.

Meyer, N., and A. Reguant-Closa. 2017. Eat as If You Could Save the Planet and Win! Sustainability Integration into Nutrition for Exercise and Sport. *Nutrients* 9(4): 412.

Mies, M. 1986. *Patriarchy and Accumulation on a World Scale*. London: Zed Books.

Monteiro, C.A., J.-C. Moubarac, R.B. Levy, and D.S. Canella. 2018. Household Availability of Ultra-Processed Foods and Obesity in Nineteen European Countries. *Public Health Nutrition* 2: 18–26.

Pearson, R. 2006. Jomon Hot Spot: Increasing Sedentism in South-Western Japan in the Incipient Jomon (14,000–9250 cal. BC) and Earliest Jomon (9250–5300 cal. BC) Periods. *World Archaeology* 38: 239–58.

Price, T.D. 2009. Ancient Farming in Eastern North America. *Proceedings of the National Academy of Sciences* 106: 6427–8.

Princen, T., M. Maniates, and K. Conca, eds. 2002. *Confronting Consumption*. Cambridge, MA: MIT Press.

Roumpani, F., and P. Hudson. 2014. The Evolution of London: The City's Near-2,000 Year History Mapped. *The Guardian*, 15 May. www.theguardian.com/cities/2014/ may/the-evolution-of-london- (Accessed 27 January 2018).

Sage, C. 2012. Commentary. "Addressing the Faustian Bargain of the Modern Food System: Connecting Sustainable Agriculture with Sustainable Consumption." *International Journal of Agricultural Sustainability* 10: 204–7.

Salleh, A., ed. 2009. *Eco-Sufficiency and Global Justice: Women Write Political Ecology*. London: Pluto Press.

Schnaiberg, A. 1980. *The Environment: From Surplus to Scarcity*. New York: Oxford University Press.

Scott, J. 2017. *Against the Grain. A Deep History of the Earliest States*. New Haven, CT: Yale University Press.

Sheehan, O., J. Watts, R.D. Gray, and Q.D. Atkinson. 2018. Coevolution of Landesque Capital Intensive Agriculture and Socio-Political Hierarchy. *Proceedings of the National Academy of Sciences* 115(14): 3628–33.

Smith, B.D. 1994. The Origins of Agriculture in the Americas. *Evolutionary Anthropology* 3: 174–84.

Standage, T. 2009. *An Edible History of Humanity*. New York: Walker & Co.

Tilly, C. 1985. War Making and State Making as Organized Crime. In, *Bringing the State Back*, edited by P. Evans, D. Rueschemeyer, and T. Skocpol, 169–87. Cambridge: Cambridge University Press.

Tilly, C. 1990. *Coercion, Capital, and European States, AD 990–1990*. Cambridge: Basil Blackwell.

Tönnies, F. 1963 [1887]. *Community and Society*. New York: Harper & Row.

UN. 2014. *World Urbanization Prospects*. New York: Department of Economic and Social Affairs, Population Division, United Nations. https://esa.un.org/unpd/wup/Publications/ Files/WUP2014-Highlights.pdf (Accessed 30 January 2018).

USBC. 2017. *US Population—Historic*. Washington DC: US Bureau of the Census, Department of Commerce. www.census.gov/ (Accessed 4 February 2018).

van der Veen, M. 2005. Gardens and Fields: The Intensity and Scale of Food Production. *World Archaeology* 37: 157–63.

Vandermeer, J., A. Aga, J. Allgeier, C. Badgley, R. Baucom, J. Blesh, L.F. Shapiro, A.D. Jones, L. Hoey, M. Jain, I. Perfecto, and M.L. Wilson. 2018. Feeding Prometheus: An Interdisciplinary Approach for Solving the Global Food Crisis. *Frontiers in Sustainable Food Systems* 2: 39. doi: 10.3389/fsufs.201800039.

Wallerstein, I. 1991. *The Politics of the World-Economy. The States, the Movements, and the Civilizations*. Cambridge: Cambridge University Press.

Wallerstein, I. 2011. *The Modern World-System: Capitalist Agriculture and the Origins of European World-Economy in the Sixteenth Century. With a New Prologue*. Berkeley: University of California Press.

Williams, R. 1973. *The Country and the City*. New York: Oxford University Press.

3

THE CHANGING CHARACTER OF THE CITY-FOOD NEXUS

In this chapter, we trace the path of the city-food relationship through the development of capitalism with particular regard for global North cities. They have garnered overwhelming attention in spite of being the least promising and most marginal for urban food production prospects. Also, we are quite familiar with several of these cities—London, New York City, and San Francisco, so it is easier for us to recount the stories of those places. This does not, in any way, mean that those cities can stand for the entire global North, much less the rest of the world. The task here is, rather, to counter the recurrent misleading claims made of urban gardening projects. Nevertheless, a broader lesson from this geographically limited sample is that urban food production prospects tend to be of marginal significance. We explore the city–food historical relationship by analysing capitalism's background structural aspects—significant but somewhat at a remove—as well as its more immediate and perceptible ones. The prime background factor bearing on the recent historical arc of urban food growing was the rise and decline of industrial production in cities of the global North. The decline was followed by urban regeneration in many cities (and urban neglect in many others) channelled through the post-industrial stage of capitalism—finance capitalism. Efforts at urban economic recovery were based in expanding a new way of securing capital for private property redevelopment: public debt. A leading example was the 1949–1973 US Urban Renewal programme (Zipp 2013).

The chapter then shifts to an analysis of urban redevelopment's relationship to the origins of the surge in urban food growing beginning in the 1970s. Our spotlight moves to foreground factors and falls on the role of gentrification. Gentrification describes a common real estate process: redeveloping a downscale neighbourhood of working-class renters so that it attracts middle-income owners by conforming to their lifestyle tastes. It is a process financed with surplus capital (from capitalist "investors" with too much cash) and regulated through urban property zoning

DOI: 10.4324/9781003131281-3

policies (Smith 1979). One case study demonstrates how the process played out in relation to the start-up of a community garden on the West Side of New York City's Manhattan Borough. Following this, we profile current contradictions of the urban food growing movement with a focus on the tension between gentrification and social justice. The chapter then outlines the global background factors which shade the movement, including increasing urbanisation and urban sprawl. We conclude with a portrayal of urban food growing's role in an intensified uneven development process initiated in the 1960s.

While the focus in this chapter is on small sections of the global North, it is important to reiterate that food growing was not abdicated in most cities of the world, especially those in the global South. In part, this is because countries in the global South did not experience the full-blown industrialisation that filled its cities with factories and their workers, though they were overwhelmed by fallout from industrial resource extraction as well as by a plethora of imported industrial consumer products. Moreover, following the first isolated and one-off accounts from Central Africa in the 1960s, the propagation of urban food growing in global South countries has become part of institutional projects, including those sponsored by the International Development Research Centre based in Ottawa, and by the UN Food and Agriculture Organisation's Development Programme based in Rome (Mougeot 1999; Smit, Nasr, and Ratta 1996).

Thus, despite the surge of urban food growing in the global North, its primary locus remains in the global South (Altieri 2012). One comparative research project found that in 11 of the 15 global South countries studied, the share of urban households growing food was over 30 per cent. Participation was concentrated among the poor, with over 50 per cent of the poorest quintile engaged in food growing in eight of the 15 countries. The urban share of national food production ranged from a low of 3 per cent in Malawi to a high of over 20 per cent in Madagascar and Nicaragua (Zezza and Tasciotti 2010). In other global South countries, the scale of urban food growing is even higher. For example, in Cuba it accounts for almost 60 per cent of all vegetable production (Premat 2005). Compared to levels of food output such as these, its present and potential production in the global North strikes one as negligible and not deserving of its promotion in status from gardening to agriculture. We devote more attention to global South cities in subsequent chapters, in part because they constitute better examples from which to draw in the quest to develop ecosocialist projects.

Cities and food in the era of globalising capitalism

The deep background (examined in Chapter 2) of historic links between food growing and cities reveals much about the character and prospects of the connection. They have been only contingently associated since the ancient origins of both. The long alliance that did happen in many locations thrived within relatively close quarters for thousands of years in various spatial formats and local sites, based in progressively larger arrays of domesticated biota and tools. Here we look at the constitution and interaction of societies through three stages of the evolving global relationship between cities and food over the last 500 years, and later turn to a focus on the global North.

Major shifts have transpired in the city–food nexus as capitalism expanded and deepened. There have been three broad developmental stages, beginning with mercantilism, continuing through industrialism into the present stage of financialism (Arrighi 1994; Gordon 1978). Of course, the organisation of food production has played a critical role in all three stages. It is also important to point out that the transitions between stages were not smooth but unpredictable and uneven—in fact, quite rocky. They featured a great deal of political struggle and military conflict within as well as among states. For example, the historical conflicts among Western Europe's long-lived (and in some respects enduring) colonial empires was not finally concluded until the end of World War II and the establishment of the UN (both in 1945) and the development of the EU, beginning with the European Coal and Steel Community under the 1951 Treaty of Paris. The complex and multifaceted developmental stages of a globalising capitalism are only highlighted here in reference to significant changes in city–food connections.

Mercantilism and food globalism

Capitalism got its first worldwide legs in commerce. The separation between people and their food sources—geographical, social, and ecological—experienced a stepwise change in the development of long-distance trade in food, a basis for the origination of the first global empires about a half-millennium ago. These empires launched the initial globalising form of capitalism—its mercantile version, or capitalism 1.0. It was preceded by regional waterborne shipping powers in the Mediterranean and other seas, as well as in lakes and navigable rivers. The stepwise change was the advent of oceanic-scale commerce stimulated by the profitable exploitation of a vast range of foods and natural resources in lands distant from Europe.

The globalising appropriation was exceedingly lucrative. Its most economically successful undertaking was the Dutch East India Company (1602–1799), a government-supported private consolidation of Dutch traders. It was the first-ever stock market-listed public company, and it is reputed to have been the richest capitalist enterprise ever. In comparative currency terms, even the contemporary digital giants—Alphabet, Amazon, and Apple—do not come close to the magnitude and duration of this company's financial success (Black 2014; Desjardins 2017). The grand prizes of the mercantile business initially were spices. New tastes to mainly upper-class Europeans featured peppercorn from the Malabar Coast, now India's Kerala state; nutmeg and clove from Java, now part of Indonesia; and cinnamon from Ceylon, now Sri Lanka (Keay 2006). While spices highlighted the start-up phase of global trading, the exploitation eventually extended across a wide range of domesticated flora. For example, the potato (*Solanum tuberosum*), now the world's fourth leading food crop (behind three cereal grains), first migrated on the ships of Spanish conquistadors from the Inka homelands of present-day Peru to Europe in the sixteenth century (Salaman 1985). The ships of colonising nations carried a wealth of new food tastes back to their European home tables from the Americas, including tomatoes, maize, and cacao (Nunn and Qian 2010).

Mercantilism's overarching motif was a genocidal settler colonialism that was spread by Western European ocean-faring nations. They built extensive global empires on the exploitation of the foods, food-growing capacities, and raw materials developed by an untold number of long-standing Indigenous communities in Africa, Asia, the Americas, Australasia, Melanesia, and Polynesia. Many of the original peoples who had lived there for millennia were eradicated or marginalised to geographic extremities and reservations through warfare, land expropriation, the destruction of resources needed for social reproduction, and the spread of settler diseases (Bodley 2015; Nunn and Qian 2010). The bulk of appropriated foods and fibres were produced on the backs of enslaved, mostly African, labourers who were traded in the new global marketplace. Perhaps the best-known human-commodity commercial arrangement involved an English-based trading triangle comprising slaves from West Africa, sugar from the Caribbean, and goods from England (Clements Library 2018).

Commodified foods, resources, and persons became a driver and main source of capital accumulation. The colonies were established in coastal settlements and then spread extensively inland. Through their rising numbers and technologies, settlers appropriated rich loads of marketable foods and resources—tea and coffee, tobacco, cotton, gold and silver, etc. This first wave of globalisation was led by, in order of per capita international trade magnitude, Holland, Portugal, England, Spain, and France (Ortiz-Ospina and Roser 2018). They warred and plundered amongst each other as well as against Indigenous peoples, dividing up whole continents and remaining in power for hundreds of years—in much of Africa, Australasia, Melanesia, Polynesia, and Asia until the mid-twentieth century. In fact, there still exist many places under such colonial dictatorship (e.g. New Caledonia, Ceuta and Melilla, Chagos, Reunion, French Guyana), aside from the descendent settler colonial regimes of the present, including all of the Americas as well as places like Guam and Palestine.

Appropriation through settler colonialism was the basis for making Western European nations the first global powers, fostering an expansion of their capital cities—Amsterdam, Lisbon, London, Paris, etc. Additionally, the process led to the establishment of coastal *entrepôt* cities (transhipment ports) in their colonies—Macao, Batavia (Jakarta), New Orleans, Singapore, among others. Kaapstad (now Capetown) was created in 1652 as a supply station for Dutch ships sailing to and from East Africa, India, and East Asia. Foods and other agricultural products, along with slaves, were the lifeblood of commercial capitalism. Meanwhile, local food growing remained integral to the daily lives of most urban dwellers in both home and *entrepôt* cities. Their farms remained close neighbours. The great bulk of the new foods arriving to the home cities were destined for the plates of royalty, aristocracy, and a small but growing trade-based bourgeoisie. Since its mercantilist origins, capitalism, built on risky speculative profiteering, has featured repetitive economic collapses, leading to widespread impoverishment. The collapse of the Dutch tulip bubble in 1634–1637 is generally recognised as the first recorded example; the 2007–2009 financial Great Recession the most recent.

Industrialism and city–food separation

A significant shift in the city–food relationship was set in motion by the advent of the industrial or manufacturing stage of capitalism—capitalism 2.0—several centuries ago, largely within the cities of capitalist polities. Mercantilism then became a steady but secondary source of capital accumulation; it expanded to include industrial products, textiles being an early example. Capitalists shifted their surpluses to investments in the exploitation of domestic working classes, often violently by means of dislocation and dispossession of peasants. The factories that came to dominate whole urban districts turned out a wide range of manufactured commodities, including new tilling, planting, and harvesting equipment that replaced many farm labourers. The trade in industrial goods followed the globalising lead of mercantilism. Mass production was extended from tools and machines to food growing itself in the twentieth century. Farming then was essentially abandoned in and near cities (except for periodic emergencies) in order to make room for a mass in-migration of factory workers, who forged and assembled a huge bounty of industrial products, including a wide range of new packaged foods.

In Victorian Britain of the late nineteenth century, tinned food rapidly grew in range and quantity to meet the needs of the expanding working classes in cities. London's population grew by over sixfold over the course of the century (DHI 2018). Tinned food had been pioneered earlier in Europe to feed armies and navies while they were abroad (Spencer 2002)—largely to expand and consolidate colonial empires. Later, in the industrial period, *Spam* (minced canned ham) appeared in the US in 1926 and beer cans in 1935 (QBV 2018). The burgeoning food industry developed gas- and electric-powered ice-making machines to replace ice boxes in order to transport its products across continents and around the world by train and ship.

At the same time, an industrial-era urban population growth of immense proportions extended the boundaries of what became global North cities, converting adjacent farms to urban sprawl, made possible by the industrial age's iconic machine—the railroad. Trains, trams, and subways provided mass transit within cities and between cities and their suburbs. Modest peri-urban farms gave way to larger ones beyond suburbs, and then to mega-scale operations located, because of their size, very far from the swelling numbers of urban-dwelling food consumers. Food, like workers, was then transported over increasing distances by wheeled machines, the energy source of which embraced petroleum in the early twentieth century.

Illustrative of the results of industrialisation, the number of farmers, as a per cent of the total labour force, fell consistently over recent centuries. Census records show that farmers in the US declined from 90 per cent of the labour force in 1790 to 43 per cent in 1890 and to a mere 3 per cent in 1990 (USDA 2017). At the same time, the number of farms fell while their size grew. Between 1910 and 1990, they declined by one-third and their average size more than trebled (Lyson 2004, 21). Today, large farms dominate food production. Most cropland is on holdings with at least 445 ha and many are five to 10 times that size. Still, farms continue to grow in size as their scale efficiencies enable them to operate more profitably than small businesses—they "utilize labor and capital more intensively, which provide them with the primary source of their financial advantage" (USDA 2013a, 6). The major share of the cheap labour they depend upon for their profits is provided by low-wage migratory farm workers, often without citizenship, especially in the largest agricultural producing state in the US—California (Alkon and Guthman 2017).

Industrial capitalism peaked in nations of the global North in the mid-twentieth century when working-class solidarity, coordinated actions (like strikes), and unionisation led to decent wages and working conditions. The numbers and densities of factory, resource extraction, and transport workers were the basis for a political shift towards domestic social democracy, illustrated, for example, by the New Deal in the US and the Welfare State in the UK. However, a transition to present-day financial capitalism has led to neoliberal austerity regimes that are devoted to a new phase of heightened inequality and labour exploitation in their quest for capital accumulation.

The post-industrial transition in the latter quarter of the twentieth century included an accelerated suburbanism that morphed into exurbanism with the assistance of automobiles and superhighways (Freund and Martin 1993). In other words, many formerly thriving inner cities were hollowed out and abandoned. Meanwhile, food increasingly became a global commodity grown in rural areas of the global North and in the countryside of the still largely agricultural global South. In the North, the rural concentrations of large-scale agricultural products became known as society's "bread baskets." In North America, they included the wheat belt of the interior Great Plains and the salad bowl of California's coastal Salinas Valley. Counterpoised to these food baskets were new urban agglomerations that featured no agriculture.

While food growing continued to develop in rural areas in ever-larger and more mechanised and chemicalised formats, many hollowed-out urban centres became targets for regeneration. It was spurred in part by the political forces brought to bear by various mid-century social movements, particularly the civil rights movement. In the first three-quarters of the twentieth century, millions of Blacks in the US had migrated from the rural South to the urban North to find jobs and escape the openly racist oppression legalised under the "Jim Crow" segregation laws. Subsequent deindustrialisation and urban decay, as well as the fact that racism was hardly limited to the South, contributed to their widespread social and political mobilisation. The densities of their populations in major cities provided a large social base needed for mass political action—just as such densities had served earlier in the mobilisation of mainly white workers into industrial unions. The US government responded to urban Black political mobilisation and struggles with bloody repression, but eventually took recourse to two major ways of countering the continuing and massive unrest resulting from reckless racial policies—creating new laws and policies to promote civil rights and expanding the financial supports for urban redevelopment programmes.

The principal US vehicle for top-down city regeneration was the Urban Renewal programme. It featured redevelopment, which transformed degraded inner city-built environments while displacing residents. A gentrification process often then changed the socioeconomic composition of neighbourhoods from a rental to an owner base, largely to the advantage of whites. The physical and demographic refashioning of communities inspired resistance. Urban Renewal acted as the bulldozer of redevelopment and was popularly referred to as slum clearance (Anderson 1964; Teaford 2010). The author James Baldwin famously called it "Negro removal" in a televised 1963 interview on WNDT-TV (New York City). Of course, redevelopment and gentrification have not been limited to deindustrialised cities in the US. The two have featured in "most of the Western advanced capitalist world" (Smith 1986, 17). They have been analysed as a general turn in capitalism towards the extraction of value from urban property (Weber 2002). It was necessarily a major undertaking in Western Europe following World War II's devastation of many of its cities. In the UK, regeneration was the preferred terminology (see Tallon 2013) and, as in the US, it was coloured with issues of race and ethnicity (Maginn 2004).

Financialism and city–food reconciliation

The urban revitalisation's economic framework was structured through an emergent corporate finance stage of capitalism—capitalism 3.0—in the global North. One of its salient characteristics was a shift to a service-based economy, in which the bulk of the labour force worked in offices, not factories. Much industrial production was transferred to rural areas of industrialised nations and to the global South, motivated by capitalists' demands for both cheap land and low-wage labour to expand factories and assembly lines (Bluestone and Harrison 1982). The successor in the present financial stage of capitalism has been the neoliberal state, the anchor policies of which are privatisation, marketisation, and austerity. The negative impacts have fallen most heavily on people of colour, with Blacks in the US being a prime example. Their resistance to the racist legacies of slavery generated a civil rights movement 2.0, prompted by the Black Lives Matter (BLM) network that emerged in 2014 and went global in 2020 (see Chapter 8).

The new financial stage in capital accumulation again shifted profit-producing activities within and between nations. For example, in the US, the leading cities of industrial production, such as Detroit (automobiles) and Pittsburgh (steel), gave way in status and prosperity to monetary transactions in New York's Wall Street. Among countries, the economic power became concentrated in multinational finance-focused corporations located in a new category of "global cities"—London, New York City, and Tokyo (Sassen 2001). This intensified concentration of wealth served to widen the already large socioeconomic gap between, as well as within, the global North and South (state-socialist systems were the most salient exception until most of them were forced to fall apart in the 1990s).

While the format for the city–food relationship has changed through the stages of capitalism's build-out, it has always remained a central component of capital accumulation. Thus, even now, in the first quarter of the twenty-first century, agricultural workers represent 28 per cent of the global labour force, with large differences between and within the North and the South. In the UK and the US, just 1 per cent of the workforce is formally employed in agriculture, but in China 25 per cent and in Indonesia 29 per cent, while in India 43 per cent are. Within Africa, there are possibly even wider margins. In Burundi 92 per cent earn wages in farming, in Nigeria 35 per cent, and in South Africa just 5 per cent (World Bank 2019). Global agricultural prices are mediated through large agricultural and mineral commodity exchange enterprises. The two largest enterprises are in New York City and in Dalian. Financial capital instruments, known as derivatives, now dominate such transactions. Their derived prices are based on the underlying values in bundled assortments of individual assets that often barely have any connection to an actually existing food item (UN 2009). They have become an icon of financial capitalism since the global market collapse of 2008.

The urban agriculture bandwagon

Beginning in the 1970s, regenerating inner city neighbourhoods in the global North began to sprout a back-to-the-future incarnation of urban food growing—including community gardens that occupied (sometimes through illegal squatting) abandoned and dishevelled lots in order to create green spaces and to grow food. At about the same time in the US, a new local-foodist movement was emerging (see Chapter 5), based in part on the 1960s hippie phenomenon, whose cultural impact had revived farm-fresh food, as well as on the emergent environmental movement that was inspiring a popularisation of organic foods. Both movements were rooted in opposition to the rise of remote large-scale agriculture and its associated chemicalisation and standardisation of food. One of the public appeals of the new movements' participants was for a return to the foods their grandparents had (grown and) eaten.

The environmentalist and local-foodist campaigns were foreground factors in an "urban renewal" of food growing. Its separation from population centres, a major trend of the twentieth century, lasted into the 1970s, when it began to give way to a post-industrial reconciliation between the now metropolitan city and locally sourced food. Data from New York State demonstrate the growth in two of the more popular means for bringing food growing and cities closer together. Urban farmers' markets and community gardens increased in number there by more than threefold over the last quarter of the twentieth century (Lyson 2004, 97). Nationally, the Community Supported Agriculture programme, supported by the US Department of Agriculture, helped to create local supply chains between smallholding farmers and city dwellers. The long distances between rural food growing and urban consumer were a motivating factor in reuniting cities and food. The average distance that food travelled from farm to consumer was by then estimated to have reached more than 2,000 km, or roughly 1,300 miles (Kelley, Harper, and Kime 2013).

The current urban gardening surge promotes another shift in the city–food relationship. It represents an attempted revival of the pre-industrial closeness and spatial integration of food and urban residence, and it is touted to increase food output of cities in the global North. However, as we show in Chapter 5, any increase will be marginal at best. Nonetheless, the bandwagon should continue because the surge presents opportunities other than growing food. What is needed is a re-direction of the bandwagon's aspirations to take advantage of the opportunities—a sort of bandwagon 2.0 (see Chapter 8).

In the early twenty-first century, urban food growing in cities of the global North has developed from a receding fringe of traditional domestic flower and vegetable gardeners into a fully loaded assortment of community organisers, environmentalists, campaigners, and entrepreneurs. This blooming of the city and food relationship has become the subject of abundant coverage by media and mounting research in academia, all of which has served as fertiliser for its growth. In industrialised countries, urban food growing, over the past century, gained such widespread public attention only when stimulated by economic crises and great wars (McClintock 2010). After World War II, urban gardens in the US "gradually disappeared from the cityscape until their rejuvenation in the mid-1970s" (Eizenberg 2013, 18). Thus, for reasons other than national emergencies, urban food growing has become a veritable hay wagon leading a parade of advocates and cheered on by an attentive public (Tortorello 2014). The procession bathes in a glow of popular optimism, sporting a (largely assumed) checklist of accomplishments: making cities greener, improving urban diets, achieving food security, and promoting food justice. In subsequent chapters, we remark on how these presumed achievements fall short. In this chapter, we concentrate on some questions that lurk beneath promotional claims about urban agriculture. What is behind this popular movement of local-foodism? What does its future look like? More importantly, what is lacking or perhaps wrong about its assumptions and declarations? What about the lack of attention to the rest of the world, where urban food production arguably has been only rarely marginal?

The contemporary urban food-growing movement in the global North has been described aptly as a big lumpy tent (Pollan 2010). The lumps represent different processes and programmes that operate at multiple levels—from the individual through the communal to the global (McClintock 2014, 165). Growing food in cities has taken a differentiated, even entropic cast in its development, as indicated in the growing use of multi-modal and multi-faceted as its descriptive terms. From a base of domestic and allotment gardening, cities around the world feature, for example, school gardens, prison gardens, entrepreneurial gardens, as well as community gardens (Ferris, Norman, and Sempik 2001; Hou 2017). In addition to the wide variations in scale from micro through macro, there are thematic and analytical oppositions or contradictions. While we use a case study method to compare its layered scales, especially as they relate to global North–South inequalities, there are macro-structural factors bearing upon these local variations. Our goal is to broaden the present framework of urban food growing by focusing on the necessity for empirically based assessments of its productivity as well as of its social and biophysical sustainability prospects.

It is important to note that while changes in the food–city relationship are highlighted largely for the US here, they have occurred around the industrialised world in a wide variety of local contexts, though very unevenly over time. The new enthusiasm for urban food growing has spurred a resurgent interest in allotments and city farms in the UK (Martin and Marsden 1999; Perez-Vasquez, Anderson, and Roger 2005). The various contours of the renewed city–food relationship have been outlined for many metropolitan areas in the global North, including Paris, London, Vancouver, and Toronto (Cockrall-King 2012). There are substantial global differences in the food-growing trend in the global North, noted here in comparisons of the UK with the US. The former has a deep legacy of shared gardening—from medieval commons through modern allotments. The development of "city farms" can be viewed as a current expression. Cities within the global North nations also demonstrate a considerable variety of urban food-growing formats. The study of three sites (Martin, Clift, and Christie 2016) illustrates this: a community farm in a suburb of London, a community garden in the heart of New York, and an agricultural park in an exurb of San Francisco.

The present urban food-growing sites have been sorted in a number of ways. One such classification has four categories: agricultural, cultural (for socialising), park (for green landscapes), and multi-purpose. The multi-purpose ones are exemplars of the community garden phenomenon in the US (Chitov 2006, 455). A standard for their success has been set by the West Side Community Garden (WSCG) in New York City. It provides an illustrative case study for this book. Later, in Chapter 5, its food output and its ecological and social sustainability are analysed along with other growing sites.

A West Side story

Landlording or real estate ownership has been a principal driver of capital accumulation in all its stages. Some 60 per cent of the global wealth today is invested in real estate (Dawson 2017). The powerful driver of land speculation in urban planning and politics has been analysed as the "real estate state" (Stein 2019). Some folks get so rich that they have too much money even for their own lavish standards and regular investments. They often decide to dump the excess cash on buying up buildings, land, and whatever else they can get ownership of, but always to turn excess capital into a money spinner. Of course, they also use the excess to buy status (philanthropy) and to bribe government officials (lobbying). These can prove decidedly useful investments towards grabbing more real estate, especially when local dwellers object to their neighbourhoods being made unaffordable (gentrification) or ransacked of public spaces. Put another way, real estate serves the wealthy as a way of unloading excess capital (which otherwise could depreciate) into speculative investments in order to continue accumulating capital. In recent decades, prosperous cities in the global North have moved from disinvestment and population decline to experiencing real estate booms (perhaps recurring bubbles-to-be) and population growth. This is the case with cities that were the sites of field investigations—in the global South as well as the global North. The path from socialised deindustrialisation ruin to privatised financialisation riches began in 1960s redevelopment and gentrification, and events in New York City illustrated the general trend.

The WSCG (Fig. 3.1) aptly illuminates the course of events that followed deindustrialisation in numerous cities of the global North (Martin 2011). The setting for these events was well described as follows:

> In the devastated urban context of the 1970s, when the first gardens appeared in New York City, community gardens were marginal spaces on the spatial maps of capitalism. They can be understood as heterotopias in the sense that they negated, and to some extent reversed, the destruction that prevailed around them. This general urban destruction of the 1970s can be seen as part of a process of . . . planned (or not) destruction of cities (or parts of cities) in order to rebuild them anew as part of a capital accumulation strategy.
>
> *(Eizenberg 2013, 22)*

The WSCG reflected this accumulation strategy as it came to ground in the redevelopment and gentrification of a physically deteriorated poor and working-class neighbourhood. The garden began informally in 1975 as a local insurgent reclamation project for a debris-strewn lot that was described as being "neck high in garbage and famous for its stripped stolen cars" (Yang 1982, C10). Today it is promoted as "The Village Green of the Upper West Side."

While a local success in its own right and terms, the WSCG also represents a material and a social confluence of major structural changes in post-industrial urban US. It is a fitting case for analysing these changes, which include economic restructuring based in finance capitalism, and the resurgence and consolidation of popular and fertile localist social movements centred on environment and food. The Garden can be seen as a contingent artefact illustrating the major storylines of how these structural changes, mediated through government agencies and real estate developers, materialised. Manhattan's Upper West Side is a better-known residential neighbourhood than most, largely because of a Broadway musical set in the mid-1950s that featured a rivalry between Puerto Rican and Irish street gangs with a Shakespearean *Romeo and Juliet* plot. The 1957 musical was followed by an equally successful 1961 film. Both were titled *West Side Story* and had, as their setting, the streets and tenements of the local area at its cusp of renewal. Following World War II, the Upper West Side was in a condition of physical decline, social disorganisation, and economic disinvestment.

FIGURE 3.1 West Side Community Garden, Manhattan, New York City, US (photo by George Martin 2012)

Certain cities and neighbourhoods comprising major public projects became the focus of the new US Urban Renewal programme; the Upper West Side serves as an exemplar. The local redevelopment scene was described by Jane Jacobs (1961, 113) as being "a badly failed area where social disintegration has been compounded by ruthless bulldozing, project building and moving people around." The most important feature of local regeneration was the construction of the Lincoln Center for the Performing Arts on a 6.6 ha site with multiple venues—for dance, music, and theatre. It was opened in 1962 on land cleared by Urban Renewal. After the Center and its northern edge were fully developed, a neighbouring area was seized by the city through eminent domain. There, "the clearance of 20 blocks of poor residential and substandard structures eliminated the threat of encroaching blight and created an initial stimulus for reinvestment" (Wilson 1987, 38). Displaced tenants replaced through gentrification produced "a new 'community' . . . composed of 'fully tax-paying' (luxury) apartments, middle-income cooperatives, rehabilitated brownstones (for middle-income families)" (Lyford 1966, 8). This 20-block neighbourhood comprises the heart of the Upper West Side, and the WSCG became one of its attracting features.

Redevelopment and gentrification can feature hard-hearted evictions as a calling card. In the WSCG neighbourhood, over 6,000 tenant households, largely lower-income Irish, Puerto Rican, and Black, were put on the streets. After the tenements were razed and new condominiums erected, its white middle-income earners became ascendant. In the Upper West Side as a whole, between 1970 and 1980, Blacks declined as a proportion of the population by 25 per cent while Whites increased by 14 per cent. In the same decade, median family income increased by 15 per cent (Wilson 1987, 42). As is frequently the case, displacement and gentrification in the Upper West Side reflected ethnic, racialised, and class inequalities.

By 2000, the Census Tract containing the WSCG had a population that was down to 8 per cent Black and 17 per cent Hispanic, considerably below the proportions for both Manhattan and New York City as a whole. The median household income was 52 per cent higher than that of Manhattan Borough and 87 per cent higher than the City's. Of the total housing units in the tract 21 per cent were built in the 1970s and 1980s, the prime decades of redevelopment that followed Urban Renewal's clearances. The former tenants of the neighbourhood's tenement buildings were evicted without compensation. They relocated to lower-income districts that were not in the innermost city and not selected for revitalisation, particularly Washington Heights and the South Bronx. The large Irish and Puerto Rican working classes featured in *West Side Story* were gone by the end of the 1980s.

The tenants did not leave without a fight. US Federal law formally mandated local citizen participation in the Urban Renewal programme. The Stryckers Bay Neighborhood Council served as an umbrella group for the block and street associations on the Upper West Side that had emerged to oppose redevelopment and its evictions. However, after considerable internal conflict, the Council eventually became a force for the White middle-income earners in the neighbourhood: "A preponderance of its white, middle-class organisational delegates wanted urban renewal to get under way as quickly as possible" (Lyford 1966, 124). A member of the Council was the 89th Street Block Association:

> On West 89th Street, block associations and the Department of Housing Preservation and Development successfully ousted squatters from a five-story abandoned building through an Appellate Court ruling. . . . The building was eventually cleared and redeveloped for market rate housing.
>
> *(Wilson 1987, 44)*

The WSCG was fashioned on this site in the mid-1970s. Wilson (1987, 41–2) remarked that "after 1973, such institutions as banks, realtors, and block associations became important in sustaining the burgeoning revitalization." Neighbourhood groups led a public campaign for an economic revitalisation, espousing "development values reflecting a concern for property values" and promoting "the importance of constructing 'aesthetics,' 'historic amenities', and 'liveable open space' in the area." This campaign gave voice to what later became known as green gentrification (see below).

Compared to public parks, community gardens proved to be quite a bargain for city governments as far as providing green spaces goes because the free labour of volunteer gardeners represents 80 per cent of the operational and maintenance investment required (Schmelzkopf 1995). Yet another gain accrues to city governments: community gardens enhance local property values, thereby adding to tax revenues (based on adding to value for property owners). In a statistical analysis of over 600 community gardens established in New York City between 1977 and 2000, Voicu and Been (2008, 268) found that:

> gardens were located on sites that acted as local disamenities within their communities. After opening, gardens have a positive impact on surrounding property values, which grow steadily over time.

The estimated net tax benefit to the City of New York over the 20-year period of community gardening emergence and expansion was an average of about $500,000 per garden (Voicu and Been 2008, 277).

At the micro level, the city government was pressed by neighbourhood residents to persuade real estate developers to reserve some land for gardens. The persuasion took advantage of the developers' interests in making (more) money. A WSCG brochure contains quotes from locals involved in the redevelopment process. According to one, "we worked to convince the developers that it was in their best interest to set aside space for a garden." A persuaded developer is quoted as saying,

> Imagine our reaction as businessmen to the idea of including a community garden on the site. But once we saw this as the best way to add to the neighborhood, the rest was easy.

In fact, the developers' position with regard to a WSCG-to-be was recalibrated several times in reaction to political pressure. Their first position was that the lot would be developed solely with market-rate housing. The second position was that it would be developed as a garden but open only to the residents of the developer's adjacent condominium complex. The third and final position was to cede the land to the city for a public garden. Thus, in the end, the speculator realised it would be a public amenity that enhanced the value of his property.

Redevelopment and gentrification are processes led by government, banks, and real estate interests, and the so-very-liberal Upper West Side was no exception. The middle-income people who moved in took advantage of a publicly subsidised opportunity to acquire a home in an increasingly desirable location featuring Lincoln Center. It was particularly appealing to a specific stratum of the post-industrial labour force—culture workers (Zukin 1987). Following the creation of the Center, the area became noted for its cultural institutions and venues. Many of its present residents work in the arts, design, entertainment, and media. Following are the estimated fractions of various labour forces represented by this clutch of occupations: For the US, one in 53; for New York City, one in 22; for Manhattan, one in nine; for the Upper West Side, one in seven; for the lead gardeners at the WSCG, one in three (Martin 2011; USCB 2005). The density of culture workers involved in the Garden calls to mind what Jane Jacobs (1961, 105) noted in her observations about who used neighbourhood parks: "the finest are stage settings for people."

In this particular case, an unusually high proportion of the people on its stage are professionals working in various capacities in cultural occupations.

Other community gardens in the global North have played a role in redevelopment and gentrification as has the WSCG. In Philadelphia, community gardening often competes with other sustainability projects in promoting social exclusion, rather than bringing improvements to oppressed communities (Rosan and Pearsall 2017). Thus, it is the case that the urban food-growing bandwagon has links with gentrification, at least in prosperous US cities. According to one anthology of analyses, a busy local food scene is a commonly recognised harbinger of gentrification (Alkon, Kato, and Sbicca 2020). Those scenes may feature community gardens as well as upscaled restaurants, cafes, and specialty grocers. As means of gentrification, a community garden can serve to "deepen societal inequities by benefitting . . . the propertied class and contributing to the displacement of lower-income households" (Horst, McClintock, and Hoey 2017, 277).

New York City community gardens are regularly targeted by private real estate developers. As noted by others, such manoeuvres do not necessarily result in the demise of gardens, due to much resistance, but some are nevertheless lost to capitalists along the way (Schmelzkopf 1995). Much of the resistance comes in the form of turning gardens into charitable (non-profit, educational) organisations or even into incorporated legal entities (e.g. limited liability companies). The Children's Magical Garden in Manhattan, as a charity promoting children's environmental education, was able to put up a relatively successful legal battle in 2014 to defend its cultivated plots from corporate attempts (Horizon Group) at clearing the garden for lucrative housing, after purchasing a lot from an absentee owner. A portion of the garden was fenced off, destroyed, and, so far, still lost to the gardeners. This was witnessed directly by one of us in 2013 who effectively lost a third of planned soil and vegetable sampling area for a research project that would have helped the gardeners assess toxicity levels. Other community gardens are razed in their entirety by city governments in order to build affordable housing (Nir 2016). There is the infamous 2006 bulldozing of South Central Farm (Los Angeles), one of the largest urban gardens in the US. The destruction—a clear case of environmental racism—was legitimated as upholding the private property rights of a speculator, to whom the city government had sold the land. San Francisco's Planning Commission displaced the city's last remaining community garden to allow a private school to be built on the site (Asimov 2016; Dineen 2016). And in Sacramento, a community garden was replaced by one restricted to residents of a newly built expensive block of flats (Cutts et al. 2017). It is ironically in the very cities where some of the headiest claims about urban agriculture are most vaunted that gardeners are forced to be ever vigilant against hostile corporate and government forces.

A political ecology contradiction of urban food growing

Community gardens represent just one relatively small lump under the big urban food tent. The tent's biggest lump is traditional home gardening. A US survey reveals that about one-third of all households are active in gardening, and of these, over nine-tenths do so at home while only about one-tenth do so in community gardens (NGA 2014). However, even if they are only a small proportion of urban food growing, community gardens are large in number. National surveys have provided the following counts: 18,000 in the US and Canada; 3,400 in Japan; and more than 1,000 in the UK (Parece and Campbell 2017, 40). Lack of data and wide variations in nomenclature are obstacles to creating comparable surveys of urban food growing in the global South. Additionally, it is not an attention-grabbing new development in cities there.

The urban agriculture bandwagon therefore evinces two broad facets in the political ecology understanding of its social justice. Findings from San Francisco as well as in New York City point to a mixed role of urban community gardens in countering gentrification, depending on gardeners' political proclivities and levels of political consciousness (Aptekar and Myers 2020; Marche 2015). On the one hand, it has provided new opportunities for residents of some poorer neighbourhoods—even, in some cases, serving to prevent or mitigate gentrification or land speculation (see the Rome case study in Chapter 7). In Pittsburgh (Gould 2019) and in Jackson, Mississippi (Akuno, Nangwaya, and Jackson 2017), self-organising Black communities have so far defended themselves successfully against encroachment by land speculators and landlords by establishing community gardens. On the other hand, food production potentially leads to elevated property values in neighbourhoods, serving effectively as a silent partner in displacement through gentrification.

Social justice

The urban agricultural parade includes stable poor, working-class, and immigrant neighbourhoods that have not been redeveloped and gentrified, and in which growing food is more than a rewarding diversion. These neighbourhoods are located outside city centres and are more common in decayed and deindustrialised locales that have not been sucked into the capitalism 3.0 world. Detroit is the pre-eminent example in the US where Black women are leading anti-consumerist struggles for local food provision (White 2011). Rosario (Argentina) offers similar examples, but for different reasons. Sustained dislocations through the combined industrialisation of farming and neoliberal policies since the 1980s led eventually to economic ruin in 2001 at the national level, sparing no city. In Rosario, the spread of hunger resulting from mass immiseration was mainly confronted by women, who began taking over unused public and private lots to produce food for their families (Ponce and Donoso 2009). There are many other similarly economically repressed, broken cities where these struggles have been developing. More than a few of them have promoted community gardening as a needed resource for deprived neighbourhoods—not as an accompaniment to gentrification. For instance, Baltimore's Cherry Hill Urban Garden has increased access to fresh foods for almost all residents of a former local food desert area (Brace et al. 2017). Not too far away, Filbert Street Garden, established in 2010, is under threat of demolition from a scandal-ridden local government's Department of Public Works (Shen 2019). What is important to note is that community gardens there are considered explicitly within a social justice framework (see Noor 2019).

The 1970s resurgence of urban food growing in the global North, especially community gardening, was often sparked by aggressive and marginally legal collective actions in poorer communities, sometimes featuring land squatting. This kind of urban food growing was not driven by city agencies, banks, and real estate speculators ("developers"). The Green Guerrillas of New York City reflect this tactic, which famously featured lobbing seed grenades over fences into vacant lots (see Hardman and Larkham 2014). In the UK, similar efforts emerged under the rubric of radical gardening (McKay 2011). There is no lack of examples in the urban agriculture surge that represent local political actions by and on behalf of neoliberalism's cast-offs.

However, the fact is that urban food growing has been primarily located in more (potentially) upscale locales such as Manhattan's Upper West Side. Illustrating its contradictions, one research concluded that "urban agriculture is more prevalent in high-income communities, but low-income communities are more likely to lack access to fresh produce, green space, and other benefits that urban agriculture can provide" (Gray, Diekmann, and Algert 2017, 33). This skewed distribution of urban food growing's benefits represents yet another variation in the old aphorism that as the rich get richer, the poor get poorer. This is because, in a capitalist system, with its exclusivist mechanisms of private property and capital accumulation, resource distribution becomes largely a zero-sum game. Those with the dough can call the shots and get more dough.

A veritable caricature of this maldistribution system is the onset of food growing in super-rich communities in metropolitan areas, perhaps illustrating a trickle-up effect—albeit a mostly stylistic one. For example, the upper-class Hamptons' villages and hamlets on Long Island outside New York have been publicised for their new bespoke vegetable gardens (Stowe 2017). Of course, the landowners do not labour, except to write cheques to the landscape architects, gardeners, and chefs who tend their domestic food output on its path from soil to plate. (And they likely have personal assistants to take care of that chore.) Gardening in itself is class-neutral but it does require access to land, which will take collective political and social mobilisation from the micro level up in order to increase the social justice potentials of its communal version. An ecosocialist perspective is one possible pathway to social justice.

Green gentrification

The idea of green gentrification is based in the growing recognition that the distribution of urban green space, especially parks, is disproportionately more available in whiter and more affluent neighbourhoods in the US (Checker 2011; Dooling 2009; Wolch, Bryne, and Newell 2014). As a result, equality of access to green spaces has become an environmental justice issue. A stream of research findings has confirmed a lack of access for racialised and ethnic minorities and the poor (Abercrombie et al. 2008; Jennings, Johnson-Gaither, and Gragg 2012; Landry and Chakraborty 2009; Wolch, Wilson, and Fehrenbach 2005). Not surprisingly, the siting of urban food-growing sites has tended to follow the discriminatory distribution of well-kept green spaces in cities. Green gentrification is a variation on greenwashing, a term developed to describe outrageous environmental claims by corporations, especially those in fossil fuel businesses. A prime example is the re-branding by British Petroleum to BP (for "Beyond Petroleum"), accompanied with a new logo: a green and yellow sunflower (Macalister and Cross 2000). It is a practice that has expanded and become more sophisticated since its introduction in the 1980s (Watson 2016). Greenwashing, in turn, is a variation on whitewashing, or the cover-up of financial crimes or scandals by means of a perfunctory and biased use of data.

A study in Vancouver found that redevelopment increasingly means green spaces to attract government and financial support. One example cited was "a developer's provision of a community garden and its soil cover" in order to hold land on speculation (Quastel 2009, 719). It is an obvious incongruity that, in the name of healthy environments and local-foods, urban agriculture can "increase inequality and thus undermine the social pillar of sustainable development" (Gould and Lewis 2017, summary on book's inside cover). The question has been raised as to whether urban revitalisation is achievable without gentrification in some variety of green dress (Green 2017). Presently, it does represent a major contradiction of urban agriculture in the global North. There is an ongoing discrepancy between, on the one hand, a progressive ideal of food justice rooted in local-foodism that aims to feed and empower communities and, on the other hand, a reality in practice that frequently excludes the ethnically and racially minoritised, and the poor.

Global backdrop to the bandwagon

In addition to the foregrounded roles played by redevelopment and gentrification, as well as environmental and local-food movements, there are more remote factors that bear upon the new wave of urban food growing in the global North. It has gained some traction from publicised concerns of scientists about several compelling global changes, including increasing urbanisation, loss of farmland (and pasture), and climate change. Food production is threatened by each, and as such they deserve more than passing consideration in assessing the assumed value of urban agriculture's output. Especially in the global South, people are seeing their homeland food sources being eliminated by the encroachment of the likes of plantation farming, mining operations, and the construction of hydroelectric dams.

The world has a rising and increasingly urban population that features the massive social inequalities and associated huge consumption differentials resulting from the globalisation and intensification of capitalist relations (as discussed in Chapter 1). By 2050, around 70 per cent of people will likely live in urban areas, compared to slightly more than 50 per cent today (UN 2015, 1976). While urbanisation and urban sprawl have peaked in the post-industrial global North, they are continuing apace in the global South. According to the FAO (2010), by 2020, 85 per cent of the poor in Latin America and 45 per cent of the poor in Africa and Asia will be concentrated in urban areas. In recent decades, China has set a new historical standard for the rate of increase as well the scale of urbanisation (Chen and Tao 2013). Globally, population and urban growth are projected to increase food demand from an expanding and (hopefully) better-fed populace by the year 2050, creating a need to produce 70 per cent more crop calories than were available in 2006 (UN 2011). Of course, for the sake of food justice and social and ecological sustainability, this will require socially and agroecologically sensible food production, more equitable food distribution, and drastically reduced food waste (now up to one-third of food produced worldwide; FAO 2019).

Urbanisation's threat to food production includes encroachment on farmland. Between 1970 and 2000, the land equivalent of Denmark was converted from farmland to urban settlement globally. The projection for 2000 to 2030 is the equivalent of Mongolia, 36 times the area of Denmark. Thus, while more food will be needed for humanity, there will be less farmland available, especially land proximate to the urban populations, where it is destined to be largely consumed (Seto et al. 2011). India, China, and many African countries have the highest rates of urban land expansion, while North America has the greatest absolute expansion in urban extent. The greatest farmland loss in terms of scale of impact is in China. Its fertile coastal plains and major river valleys are home to both its largest urban agglomerations, which have been expanding rapidly, and its greatest agricultural outputs. For example, Chongqing (see Chapter 7) lays inland at the confluence of the Yangtze and Jialing rivers, and Shanghai is on the Yangtze's delta at the East China Sea. Both areas are historically rich in food output, and both have grown rapidly in recent decades, joining the list of the world's largest urban agglomerations. To illustrate the scale of China's farm loss, in 2003 alone more than 2 per cent of the country's agricultural land was lost to urban expansion (Chen 2007; Martin 2007).

On top of a general disappearance of farmland due to urban sprawl, climate change is projected to result in a farm yield loss, making food production shortfalls among the principal threats for humanity (FAO 2013; IPCC 2016). Among other problems, climate change is forecast to make for longer and deeper droughts that will alter the volumes, timing, and distribution of water resources. In 2017, 70 per cent of global freshwater was used in agriculture (Khokhar 2017). In many cities and countrysides, new water demands will exceed surface water availability, creating "a high potential for conflict between urban and agricultural sectors" (Florke, Schneider, and McDonald 2018, 51). Freshwater has become a terrain contested between interests: public access versus neoliberal policies that, since the 1970s, have promoted capital accumulation through various tactics of theft (Swyngedouw 2006). Such conflict over access to water has been common in California (e.g. Reisner 1986), the greatest food-producing state in a country that is the world's third largest food producer and largest food exporter and that is expected to see a decline in yield for major crops by mid-century due to rising temperature and precipitation extremes (USDA 2013b). Conflicts over water in capitalist systems are generally likely to escalate, some perhaps quite steeply. Moreover, California has been the scene of ever-more destructive wildfires due to the extreme temperatures and dry spells produced by global warming (UCS 2018). They represent a growing threat to forests and pasture lands, as well as to human settlements.

As a consequence of these unfavourable world climate and demographic trends, global agriculture faces the daunting dual challenge of increasing production levels substantially and doing so equitably and sustainably in social and ecological terms. Capitalist relations worldwide have historically contributed to this dual challenge through an economy based on fossil fuel use, massive land speculation leading to urban sprawl, and many other routine business activities that wreak social and ecological havoc (Dawson 2017). Designating the upsurge in urban food growing as agriculture indicates an assumption that increasing the output in cities will contribute significantly to meeting this capitalism-induced dual challenge. We challenge that assumption in Chapter 5. In fact, we argue that urban agriculture may serve to deflect needed attention away from more useful and efficient approaches to addressing the food and ecological threats that lay ahead for humanity, including the redistribution of food according to need, not profitability.

Local political ecologies

There is a new local-food politics driving the bandwagon of rising interest in urban food growing in the global North, based in the environmental movement and related campaigns for organic, locally sourced (fresh), healthy and sustainable diets. However, examples of this phenomenon can also be seen in places in the global South, including Shanghai, Havana, and Beijing. All the locales of our case study research underlying this book illustrate this movement at work—in Chongqing, London, New York City, Rome, and San Francisco. Contemporary community food growing evokes a cultural orientation different from that of traditional urban domestic gardens (and, with exceptions, allotment gardens). It arose as neighbourhood mobilisations to redevelop vacant and derelict lots in post-industrial cities of the global North. However, in other parts of the world, such as East Asia, the process may involve a combination of neighbourhood-level actions motivated by deprivation and by farming lifestyle continuities among displaced people (who may be overtaken by urban expansion), as discussed in Chapter 7.

Scholars and activists are questioning the presumption that farms and gardens have only positive or liberating functions (Reynolds and Cohen 2016, 6). Pudup (2008), for example, has proposed that community gardens (which she identifies as organised garden projects) have been used to cultivate citizen-subjects who may act either in step with or in opposition to an austerity-minded neoliberal state. Still others have argued that for urban agriculture to lead to structural change, it must simultaneously and contradictorily be radical in approach *and* engage actively with the mainstream capitalist market system (McClintock, Cooper, and Khandeshi 2013; Bródy and de Wilde 2020). These may be among the reasons why urban agriculture has been found to mask deeper structural inequities (Colasanti, Litjens, and Hamm 2012; Cohen and Reynolds 2014; DeLind 2011; Yakini 2013). In the end, as one researcher concluded, "overemphasizing the benefits of urban agriculture without regard to its downsides is dangerous and risks marginalizing this movement" (Mok et al. 2014, 38).

The rising interest in urban food production should therefore be understood as originating from multiple kinds of dynamics that are locality-specific yet connected by global capitalist inter-linkages. Unlike long pre-existing modes of producing food in cities, more recent initiatives originate in variegated grassroots and institutional movements often motivated by environmental concerns and without much, if any, food growing experience, not just in global North cities, but also in the South. Overall, involvement is rarely kinship-based in such cities, and there is a tendency to privilege organic vegetable cultivation, though not necessarily out of any environmental convictions (see, for example, Bretzel et al. 2016).

Despite general trends there is a need to be mindful of context. In some instances in the global North, urban food production is spurred by chronically high unemployment and declining wages as in failed cities, such as Detroit, which have been devastated by deindustrialisation followed by land speculation and state-supported corporate theft (also known as the "financial crisis") since the early 2000s era of neoliberal austerity. Detroit has led the decline of US Midwestern industrial belt cities; it has been brutal there (Rose 2016, 175–7; White 2011). Its population declined by nearly two-thirds between 1950 and 2020. The outcome may lead to a harkening back to past episodes of privation-induced vegetable gardening (like the Victory Gardens during World Wars I and II), but in those days it was, crucially, government-supported, rather than a result of disastrous government policies. Such current food-growing situations resemble what has been happening in Athens and Lisbon (Ioannou, Moran, and Sondermann 2016) or Alicante and Dublin (Corcoran, Kettle, and O'Callaghan 2017; Espinosa Seguí, Maćkiewicz, and Rosol 2017) in the global North, as well as in Chongqing and other places in the global South, but for substantively different underlying reasons there. In Paris, as in Rome, investment pressures largely influence the fate of urban food production, but squatters have become more prominent in engaging with the sort of urban gardening that directly questions capitalist relations (Demailly and Darly 2017; Mudu and Marini 2018). In Budapest, Portland, Vancouver, and Vienna, urban gardening is pushed from above as part of wider urban sustainability schemes (Bársony, Lengyel, and Perpék 2019; Bársony 2020; Bende 2016; Darly and McClintock 2017; McClintock 2018; Kumnig 2017). In Prague and Warsaw, economic downturns and/or major occupational shifts among workers have undermined allotment gardens and intensified reliance on buying food (Bartłomiejski and Kowalewski 2019; Spilková and Vágner 2018).

These are some of the issues that are glossed over by mainstream urban agriculture enthusiasts. However, as many have shown, the bandwagon is proving to have some squeaky wheels. Some go further to argue that it supports neoliberalism by filling in the inequality gaps bequeathed from the privatisation of state social functions (Heynen, Kurtz, and Trauger 2012; McClintock 2014; Rosol 2010; Tornaghi 2014; Weissman 2015a, 2015b). Others point to the racialised and class inequalities in urban agriculture; in effect, a primarily White and bourgeois undertaking (Cohen, Reynolds, and Sanghvi 2012; Crouch 2012; McClintock, Cooper, and Khandeshi 2013; Meenar and Hoover 2012; Metcalf and Widener 2011) or represented as such when it is not (Reynolds and Cohen 2016). There is a compelling need to be attentive to this emerging body of constructive criticism as well as of the context-specificity and convergences that exist under the overarching rubric of urban food growing.

Conclusion

Redevelopment and gentrification are based on a finance capitalist economic calculus. A Marxist analysis by Neil Smith (1982, 139) posits that the two comprise "the leading edge of a larger process of uneven development which is a specific process rooted in the capitalist mode of production." It becomes more apparent in times of economic crisis, such as existed in the post-industrial global North of the 1970s. If successful, uneven development serves as a lever for preserving capital wealth by financial recovery via new profit-mining techniques in specific sectors and locales.

Urban redevelopment accompanied by gentrification was decidedly an uneven phenomenon in the last quarter of the twentieth century. Urban neglect was common but did not inspire attention. In 1960s Urban Renewal programmes in the US, many cities in its "rust belt" (Mahaney 2017) were never renewed or were so for merely cosmetic reasons. There was also uneven development within the urban renewed cities, as poor and working-class neighbourhoods not strategically (speculation-wise) located were left untouched. On a macro level, uneven redevelopment programmes such as Urban Renewal contributed to the growing budgetary deficits of governments—their fiscal crises then became permanent (O'Connor 1973). Financial capitalism's uneven economic recovery has been followed in the present era by the neoliberal state and its austerity and deregulation tactics—ostensibly to reduce public debt, but, in reality, to protect private capital accumulation. Thus far, the major outcomes have been growing socioeconomic inequalities, anti-immigrant populist nationalisms, and intensified political conflicts. It can be argued that the UK's Brexit and the US's Trump (both dating from 2016) reflect these outcomes.

Ultimately, for urban food growing in the global North to continue expanding in a just direction, we argue that it will need to move towards a twin focus on social and political ecological issues and away from a devotion to food production. Also, it will need to encourage mobilisations that challenge the inequitable bourgeois face of local-foodism as well as its shoring up of neoliberalism's increasing socioeconomic inequalities. Research indicates urban food production cannot reach the level of agriculture in terms of productivity, and its ecological sustainability is questionable (see Chapter 5). Its future will be determined in part by the interplay between its bandwagon supporters and their critics—critics who as yet do not present systematic alternatives to existing urban food growing. Meanwhile, in the cities of the global South, where urban food growing is most practised and developed (and often most needed, as also in the global North's devastated city neighbourhoods), there may be a levelling off in food production as cities are rapidly expanding. This convergence from disparate directions requires its own research as to the locally specific ecological and output potentials of urban food growing. Presently, the biophysical (ecological and physical environmental) aspects of food production are a recurring blind spot in many of the writings on these topics. In the next chapter, we take up the major biophysical processes affecting food production in cities by showing how urban contexts are simultaneously ecosystems. We argue, unlike current ecological takes on the city, that a historical and materialist dialectical approach to cities as ecosystems is necessary not only to comprehend the dynamics of urban food production but also to distil strategies that can contribute effectively to ecosocialist ends.

References

Abercrombie, L., J. Sallis, T. Conway, L. Frank, B. Salens, and J. Chapman. 2008. Income and Racial Disparities in Access to Public Parks and Private Recreation Facilities. *American Journal of Preventive Medicine* 34: 9–15.

Akuno, K., A. Nangwaya, and C. Jackson, eds. 2017. *Jackson Rising. The Struggle for Economic Democracy and Black Self-Determination in Jackson, Mississippi.* Jackson: Daraja Press.

Alkon, A., and J. Guthman, eds. 2017. *The New Food Activism: Opposition, Cooperation, and Collective Action.* Berkeley: University of California Press.

Alkon, A., Y. Kato, and J. Sbicca. 2020. *A Recipe for Gentrification: Food, Power, and Resistance in the City.* New York: New York University Press.

Altieri, M. 2012. The Scaling Up of Agroecology: Spreading the Hope for Food Sovereignty and Resiliency. Rio de Janeiro: UN Conference on Sustainable Development, Rio+20. May. *Sociedad Científica Latinoamericana de Agroecología (SOCLA).* www.argoeco.org/socla (Accessed 4 February 2018).

Anderson, M. 1964. *The Federal Bulldozer: A Critical Analysis of Urban Renewal, 1949–1962.* Cambridge MA: MIT Press.

Aptekar, S., and J.S. Myers. 2020. The Tale of Two Community Gardens: Green Aesthetics Versus Food Justice in the Big Apple. *Agriculture and Human Values* 37: 779–92.

Arrighi, G. 1994. *The Long Twentieth Century: Money, Power and the Origins of Our Times.* London: Verso.

Asimov, N. 2016. Panel Unanimously Backs School on Site of Urban Farm. *San Francisco Chronicle*, 30 September: 7.

Bársony, F. 2020. Városi Közösségi Kertek Magyarországon. [Urban community allotment gardens in Hungary] *Tér és Társadalom* 34(1): 1–20.

Bársony, F., G. Lengyel, and E. Perpék. 2019. Enclave Deliberation and Common-Pool Resources: An Attempt to Apply Civic Preference Forum on Community Gardening in Hungary. *Quality and Quantity* 54: 687–708.

Bartłomiejski, R., and M. Kowalewski. 2019. Polish Urban Allotment Gardens as 'Slow City' Enclaves. *Sustainability* 11: 3228.

Bende, C. 2016. A Közösségi Kertek, mint a Nagyvárosi Dzsentrikikációs Folyamatok Produktumai? A Budapesti Leonardo Eseste. [Community allotment gardens as the products of large-city gentrification processes? The case of Leonardo Eseste in Budapest] *Településföldrajzi Tanulmányok* 2: 38–52.

Black, S. 2014. This Was the Most Valuable Company in History (Worth 10 Times as Much as Apple). www.soverignman.com/ (Accessed 30 July 2018).

Bluestone, B., and B. Harrison. 1982. *The Deindustrialization of America*. New York: Basic Books.

Bodley, J.M. 2015. *Victims of Progress*, 6th ed. Lanham: Rowman & Littlefield.

Brace, A., N. Braunstein, B. Finkelstein, and D. Beall. 2017. Promoting Healthy Food Access in an Urban Food Desert in a Baltimore City Neighborhood. *Food Studies: An Interdisciplinary Journal* 7: 17–30.

Bretzel, F., M. Calderisi, M. Scatena, and R. Pini. 2016. Soil Quality Is Key for Planning and Managing Urban Allotments Intended for the Sustainable Production of Home-Consumption Vegetables. *Environmental Science and Pollution Research* 23: 17753–60.

Bródy, L.S., and M. de Wilde. 2020. Cultivating Food or Cultivating Citizens? On the Governance and Potential of Community Gardens in Amsterdam. *Local Environment* 25(3): 243–57.

Checker, M. 2011. Wiped Out by the 'Greenwave': Environmental Gentrification and the Paradoxical Politics of Urban Sustainability. *City & Society* 23: 210–29.

Chen, J. 2007. Rapid Urbanization in China: A Real Challenge to Soil Protection and Food Security. *Catena* 69: 1–15.

Chen, M., and X. Tao. 2013. Evolution and Assessment on China's Urbanization 1960–2010: Under-Urbanization or Over-Urbanization? *Habitat International* 38: 25–33.

Chitov, D. 2006. Cultivating Social Capital on Urban Plots: Community Gardens in New York City. *Humanity and Society* 30: 437–62.

Clements Library. 2018. *Sugar and Slavery*. Ann Arbor: Clements Library Collections, University of Michigan. https://theclementslibrary.umich.edu/ (Accessed 30 July 2018).

Cockrall-King, J. 2012. *Food and the City: Urban Agriculture and the New Food Revolution*. Amherst NY: Prometheus.

Cohen, N., and K. Reynolds. 2014. Resource Needs for a Socially Just and Sustainable Urban Agriculture System: Lessons from New York City. *Renewable Agriculture and Food Systems* 30: 103–14.

Cohen, N., K. Reynolds, and R. Sanghvi. 2012. *Five Borough Farm: Seeding the Future of Urban Agriculture in New York City*. New York: Design Trust for Public Space.

Colasanti, K., C. Litjens, and M. Hamm. 2012. *Growing Food in the City: The Production Potential of Detroit's Vacant Land*. East Lansing: The CS Mott Group for Sustainable Food Systems, Michigan State University.

Corcoran, M., P. Kettle, and C. O'Callaghan. 2017. Green Shoots in Vacant Plots? Urban Agriculture and Austerity in Post-Crash Ireland. *ACME: An International Journal for Critical Geographies* 16: 305–31.

Crouch, P. 2012. Evolution or Gentrification: Do Urban Farms Lead to Higher Rents? *Grist*. http://grist.org/food/evolution-or-gentrification-do-urban-farms-lead-tohigher-rents/ (Accessed 15 September 2017).

Cutts, B.B., J.K. London, S. Meiners, K. Schwarz, and M.L. Cadenasso. 2017. Moving Dirt: Soil, Lead, and the Dynamic Spatial Politics of Urban Gardening. *Local Environment* 22: 998–1018.

Darly, S., and N. McClintock. 2017. Introduction to Urban Agriculture in the Neoliberal City: Critical European Perspectives. *ACME: An International Journal for Critical Geographies* 16: 224–31.

Dawson, A. 2017. *Extreme Cities: The Peril and Promise of Urban Life in the Age of Climate Change*. London: Verso.

DeLind, L. 2011. Are Local Food and the Local Food Movement Taking Us Where We Want to Go? Or Are We Hitching Our Wagons to the Wrong Stars? *Agriculture and Human Values* 28: 273–83.

Demailly, K.-E., and S. Darly. 2017. Urban Agriculture on the Move in Paris: The Routes of Temporary Gardening in the Neoliberal City. *ACME: An International Journal for Critical Geographies* 16: 332–61.

Desjardins, J. 2017. The Most Valuable Companies of All-Time. www.visualcapitalist.com/ (Accessed 30 July 2018).

DHI. 2018. *A Population History of London 1800–1913—Proceedings of the Old Bailey*. Sheffield: Old Bailey Proceedings Online, Digital Humanities Institute, University of Sheffield. www.oldbaileyonline.org/Londonpopulation/ (Accessed 30 July 2018).

Dineen, J.K. 2016. Last Farm in S.F. May Give Way to Private School. *San Francisco Chronicle*, September 29: D1, D2.

Dooling, S. 2009. Ecological Gentrification: A Research Agenda Exploring Justice in the City. *International Journal of Urban and Regional Research* 33: 621–39.

Eizenberg, E. 2013. *From the Ground Up: Community Gardens in New York City and the Politics of Spatial Transformation*. Farnham: Ashgate.

Espinosa Seguí, A., B. Maćkiewicz, and M. Rosol. 2017. From Leisure to Necessity: Urban Allotments in Alicante Province, Spain, in Times of Crisis. *ACME: An International Journal for Critical Geographies* 16: 276–304.

FAO. 2010. *Growing Greener Cities*. Rome: Food and Agriculture Organization, United Nations. www.fao.org/ag/agp/greenercities/pdf/GGC-en.pdf (Accessed 14 January 2017).

FAO. 2013. *Statistical Yearbook*. Rome: Food and Agriculture Organization, United Nations.

FAO. 2019. *Food Loss and Food Waste*. www.fao.org/food-loss-and-food-waste/en/ (Accessed 16 May 2019).

Ferris, J., C. Norman, and J. Sempik. 2001. People, Land and Sustainability: Community Gardens and the Social Dimension of Sustainable Development. *Social Policy & Administration* 35: 559–68.

Florke, M., C. Schneider, and R. McDonald. 2018. Water Competition Between Cities and Agriculture Driven by Climate Change and Urban Growth. *Nature Sustainability* 1: 51–8.

Freund, P., and G. Martin. 1993. *The Ecology of the Automobile*. Montreal: Black Rose.

Gordon, D. 1978. Capitalist Development and the History of American Cities. In, *Marxism and the Metropolis: New Perspectives in Urban Political Economy*, edited by W. Tabb, and L. Sawers, 25–63. New York: Oxford University Press.

Gould, K.A. 2019. Pittsburgh Grows Urban Gardens in the Fight against Gentrification. https://civileats.com/2019/12/06/pittsburgh-grows-urban-gardens-in-the-fight-against-gentrification/ (Accessed 13 May 2020).

Gould, K.A., and T.L. Lewis. 2017. *Green Gentrification: Urban Sustainability and the Struggle for Environmental Justice*. Abingdon: Routledge.

Gray, L., L. Diekmann, and S. Algert. 2017. North American Urban Agriculture: Barriers and Benefits. In, *Global Urban Agriculture*, edited by A.M.G.A. WinklerPrins, 24–37. Boston, MA: CABI.

Green, J. 2017. Is Urban Revitalization without Gentrification Possible? *The Dirt*, 21 October. https://dirt.asla.org/ (Accessed 5 April 2018).

Hardman, M., and P.J. Larkham. 2014. *Informal Urban Agriculture: The Secret Lives of Guerrilla Gardeners*. New York and London: Springer.

Heynen, N., H.E. Kurtz, and A. Trauger. 2012. Food Justice, Hunger and the City. *Geography Compass* 6: 304–11.

Horst, M., N. McClintock, and L. Hoey. 2017. The Intersection of Planning, Urban Agriculture, and Food Justice. Review Essay. *Journal of the American Planning Association* 83(3): 277–95.

Hou, Jeffrey. 2017. Urban Community Gardens as Multimodal Social Spaces. In, *Greening Cities: Forms and Functions*, edited by P.Y. Tan, and C.Y. Jim, Ch. 6. Singapore: Springer.

Ioannou, B., M. Moran, and M. Sondermann. 2016. Grassroots Gardening Movement: Towards Cooperative Forms of Green Urban Development. In, *Urban Allotment Gardens in Europe*, edited by S. Bell, R. Fox-Kämper, N. Keshavarz, M. Benson, S. Caputo, S. Noori, and A. Voigt, 84–112. London: Routledge and Earthscan.

IPCC. 2016. *Intergovernmental Panel on Climate Change*, Report of Working Group II: Impacts, Adaptation and Vulnerability, Fifth Assessment Report. www.ipcc.ch/ report/ ar5/wg2/ (Accessed 25 March 2018).

Jacobs, J. 1961. *The Death and Life of Great American Cities*. New York: Random House.

Jennings, V., C. Johnson-Gaither, and R. Gragg. 2012. Promoting Environmental Justice Through Urban Green Space Access. *Environmental Justice* 5: 1–7.

Keay, J. 2006. *The Spice Route A History*. Berkeley: University of California Press.

Kelley, K., J.K. Harper, and L. Kime. 2013. *Community Supported Agriculture*. University Park PA: Penn State Extension. www.extension.psu.edu/community-supported-agriculture (Accessed 29 July 2018).

Khokhar, T. 2017. *Chart: Globally, 70% of Freshwater Is Used for Agriculture*. Washington DC: The Data Blog, World Bank. www.blogs.worldbank.org/Freshwater-/ (Accessed 1 August 2018).

Kumnig, S. 2017. Between Green Image Production, Participatory Politics and Growth: Urban Agriculture and Gardens in the Context of Neoliberal Urban Development in Vienna. *ACME: An International Journal for Critical Geographies* 16: 232–48.

Landry, S., and J. Chakraborty. 2009. Street Trees and Equity: Evaluating the Spatial Distribution of an Urban Amenity. *Environment and Planning A* 41: 2651–70.

Lyford, J.P. 1966. *The Airtight Cage: A Study of New York's West Side*. New York: Harper & Row.

Lyson, T.A. 2004. *Civic Agriculture: Reconnecting Farm, Food, and Community*. Medford MA: Tufts University Press.

Macalister, T., and E. Cross. 2000. BP Rebrands on a Global Scale. *The Guardian*, 24 July. https://theguardian.com/busines/2000/jul/25/bp (Accessed 29 July 2020).

Maginn, P.J. 2004. *Urban Regeneration, Community Power and the (In)Significance of Race*. London: Routledge.

Mahaney, E. 2017. A Geographic Overview of the Rust Belt. *ThoughtCo*, 23 August. www.thoughtco.com/rust-belt-industrial-heartland-of-the-united-states-1435759 (Accessed 1 August 2018).

Marche, G. 2015. What Can Urban Gardening Really Do about Gentrification? A Case-Study of Three San Francisco Community Gardens. *European Journal of American Studies* 10(3). https://journals.openedition.org/ejas/11316#citedby (Accessed 13 May 2020).

Martin, G. 2007. Global Motorization, Social Ecology and China. *Area* 39: 66–73.

Martin, G. 2011. *Transforming a Derelict Public Property into a Vibrant Public Space: The Case of Manhattan's West Side Community Garden*. London: Royal Geographical Society Annual Conference.

Martin, G., R. Clift, and I. Christie. 2016. Urban Cultivation and Its Contributions to Sustainability: Nibbles of Food but Oodles of Social Capital. *Sustainability* 8(5): 409.

Martin, R., and T. Marsden. 1999. Food for Urban Spaces: The Development of Urban Food Production in England and Wales. *International Planning Studies* 4: 389–42.

McClintock, N. 2010. Why Farm the City? Theorizing Urban Agriculture Through a Lens of Metabolic Rift. *Cambridge Journal of Regions, Economy and Society* 3: 191–207.

McClintock, N. 2014. Radical, Reformist, and Garden-Variety Neoliberal: Coming to Terms with Urban Agriculture's Contradictions. *Local Environment* 19: 147–71.

McClintock, N. 2018. Cultivating (a) Sustainability Capital: Urban Agriculture, Ecogentrification, and the Uneven Valorization of Social Reproduction. *Annals of the American Association of Geographers* 108: 579–90.

McClintock, N., J. Cooper, and S. Khandeshi. 2013. Assessing the Potential Contribution of Vacant Land to Urban Vegetable Production and Consumption in Oakland, California. *Landscape and Urban Planning* 111: 46–58.

McKay, G. 2011. *Radical Gardening: Politics, Idealism & Rebellion in the Garden*. London: Frances Lincoln.

Meenar, M., and B. Hoover. 2012. Community Food Security via Urban Agriculture: Understanding People, Place, Economy, and Accessibility from a Food Justice Perspective. *Journal of Agriculture, Food Systems, and Community Development* 3: 1–18.

Metcalf, S.S., and M.J. Widener. 2011. Growing Buffalo's Capacity for Local Food: A Systems Framework for Sustainable Agriculture. *Applied Geography* 31: 1242–51.

Mok, H.-F., V. Williamson, J. Grove, K. Burry, F. Barker, and A. Hamilton. 2014. Strawberry Fields Forever? Urban Agriculture in Developed Countries: A Review. *Agronomy for Sustainable Development* 34: 21–43.

Mougeot, L. 1999. Urban Agriculture: Definition, Presence, Potentials and Risks, and Policy Challenges. International Workshop on Growing Cities Growing Food: Urban Agriculture on the Policy Agenda, Havana, October 11–15. https://idl-bnc-idrc.dspacedirect.org/ bitstream/handle/10625/26429/117785.pdf (Accessed 28 March 2018).

Mudu, P., and A. Marini. 2018. Radical Urban Horticulture for Food Autonomy: Beyond the Community Gardens Experience. *Antipode* 50: 549–73.

NGA. 2014. *Garden to Table: A 5-year Look at Food Gardening in America*. South Burlington VT: National Gardening Association. http://garden.org/special/pdf/2014-NGA-Garden-to-Table.pdf (Accessed 5 February 2018).

Nir, S.M. 2016. Community Gardens Imperiled by Plans for Affordable Housing. *The New York Times*, 16 January: A14.

Noor, D. 2019. Community Rallies to Defend Community Garden. *The Real News Network*, 22 March. https://therealnews.com/stories/community-rallies-to-defend-community-garden (Accessed 30 July 2020).

Nunn, N., and N. Qian. 2010. The Columbian Exchange: A History of Disease, Food, and Ideas. *Journal of Economic Perspectives* 24: 163–88.

O'Connor, J. 1973. *The Fiscal Crisis of the State*. New York: St. Martin's Press.

Ortiz-Ospina, E., and M. Roser. 2018. International Trade. *OurWorldinData*. https://ourworldindata.org/international-trade (Accessed 26 July 2018).

Parece, T.E., and J.B. Campbell. 2017. A Survey of Urban Community Gardeners in the USA. In, *Global Urban Agriculture*, edited by A.M.G.A. WinklerPrins, 38–49. Boston, MA: CABI.

Perez-Vasquez, A., S. Anderson, and A. Roger. 2005. Assessing Benefits from Allotments as a Component of Urban Agriculture in England. In, *Agropolis: The Social, Political and Environmental Dimensions of Urban Agriculture*, edited by L. Mongeot, 240–66. London: Earthscan/Routledge.

Pollan, M. 2010. The Food Movement, Rising. *New York Review of Books* 10: 31–3.
Ponce, M., and L. Donoso. 2009. Urban Agriculture as a Strategy to Promote Equality of Opportunities and Rights for Men and Women in Rosario, Argentina. In, *Women Feeding Cities: Mainstreaming Gender in Urban Agriculture and Food Security*, edited by A. Hovorka, H. de Zeeuw, and M. Njenga, 157–66. Warwickshire: Practical Action Publishing.
Premat, A. 2005. Moving between the Plan and the Ground: Shifting Perspectives on Urban Agriculture in Havana, Cuba. In, *Agropolis: The Social, Political and Environmental Dimensions of Urban Agriculture*, edited by L. Mongeot, 153–85. London: Earthscan/Routledge.
Pudup, M.B. 2008. It Takes a Garden: Cultivating Citizen-Subjects in Organized Garden Projects. *Geoforum* 39: 1228–40.
QBV. 2018. History of the Can. *Quality by Vision*. www.qbyv.com/product/history-of-the-can (Accessed 12 January 2018).
Quastel, N. 2009. Political Ecologies of Gentrification. *Urban Geography* 30: 694–725.
Reisner, M. 1986. *Cadillac Desert: The American West and Its Disappearing Water*. New York: Viking Penguin.
Reynolds, K., and N. Cohen, eds. 2016. *Beyond the Kale. Urban Agriculture and Social Justice Activism in New York City*. Athens: University of Georgia Press.
Rosan, C.D., and H. Pearsall. 2017. *Growing a Sustainable City? The Question of Urban Agriculture*. Toronto: University of Toronto Press.
Rose, J. 2016. *The Well-Tempered City*. New York: Harper Collins.
Rosol, M. 2010. Public Participation in Post-Fordist Urban Green Space Governance: The Case of Community Gardens in Berlin. *International Journal of Urban and Regional Research* 34: 548–63.
Salaman, R.N. 1985. *The History and Social Influence of the Potato*. Cambridge: Cambridge University Press.
Sassen, S. 2001. *The Global City*. Princeton, NJ: Princeton University Press.
Schmelzkopf, K. 1995. Urban Community Gardens as a Contested Space. *Geographical Review* 85: 364–80.
Seto, K.M., B. Fragkais, M. Guneralp, and M. Reilly. 2011. A Meta-Analysis of Global Urban Land Expansion. *PLoS One* 6. doi: 10.1371/journal.pone.002377.
Shen, F. 2019. Community Demands Chow Commit to Saving Filbert Street Garden. *Baltimore Brew*, 16 May. www.baltimorebrew.com/2019/05/12/community-demands-chow-commit-to-saving-filbert-street-garden/ (Accessed 30 July 2020).
Smit, J., J. Nasr, and A. Ratta. 1996. *Urban Agriculture: Food, Jobs and Sustainable Cities*. New York: United Nations Development Programme.
Smith, N. 1979. Toward a Theory of Gentrification: A Back to the City Movement by Capital, not People. *Journal of the American Planning Association* 45: 538–48.
Smith, N. 1982. Gentrification and Uneven Development. *Economic Geography* 58: 139–55.
Smith, N. 1986. Gentrification, the Frontier, and the Restructuring of Urban Space. In, *Gentrification of the City*, edited by N. Smith, and P. Williams, 15–34. Boston, MA: Allen & Unwin.
Spencer, C. 2002. *British Food: An Extraordinary Thousand Years of History*. New York: Columbia University Press.
Spilková, J., and J. Vágner. 2018. Food Gardens as Important Elements of Urban Agriculture: Spatio-Developmental Trends and Future Prospects for Urban Gardening in Czechia. *Norsk Geografisk Tidsskrift—Norwegian Journal of Geography* 72: 1–12.
Stein, S. 2019. *Capital City: Gentrification and the Real Estate State*. London and New York: Verso.
Stowe, S. 2017. Gardening with a Checkbook: On Long Island's East End, Growing Vegetables Can Be an Expensive Habit. *The New York Times*, 6 September: D1, D7.

Swyngedouw, E. 2006. Dispossessing H_2O: The Contested Terrain of Water Privatization. *Capitalism Nature Socialism* 16: 81–98.
Tallon, A. 2013. *Urban Regeneration in the UK*. London: Routledge.
Teaford, J.C. 2010. Urban Renewal and Its Aftermath. *Housing Policy Debate* 11: 443–65.
Tornaghi, C. 2014. Critical Geography of Urban Agriculture. *Progress in Human Geography* 38: 551–67.
Tortorello, M. 2014. Mother Nature's Daughters. *The New York Times*, 28 August: D1.
UCS. 2018. *Is Global Warming Fueling Increased Wildfire Risks*? Cambridge MA: Union of Concerned Scientists. www.ucsusa.org/global-warming/science-and-impacts/impacts/global-warming-and-wildlife.html#.W2tSMyhKjak (Accessed 8 August 2018).
UN. 1976. *Orders of Magnitude of the World's Urban Population in History*. New York: Population Division, Department of Economic and Social Affairs, United Nations. https://esa.un.org/unpd/wup/ArchiveFiles/studies/ (Accessed 6 April 2018).
UN. 2009. *Overview of the World's Commodity Exchanges—2017*. New York and Geneva: United Nations Conference on Trade and Development.
UN. 2011. *World Population Prospects: The 2010 Revision*. New York: Department of Social and Economic Affairs, United Nations.
UN. 2015. *World Urbanization Prospects. The 2014 Revision*. New York. Department of Social and Economic Affairs, United Nations. https://esa.un.org/unpd/wup/Publications/.
USCB. 2005. *The 2000 National Census*. Washington, DC: Census Bureau, US Department of Commerce.
USDA. 2013a. *Farm Size and the Organization of U.S. Crop Farming*. Washington, DC: US Department of Agriculture. www.ers.usda.gov (Accessed 4 April 2018).
USDA. 2013b. *Climate Change and Agriculture in the United States: Effects and Adaptation*. Technical Bulletin 1935. Washington, DC: US Department of Agriculture.
USDA. 2017. *National Agriculture Statistics*. Washington, DC: US Department of Agriculture. www.usda.mannlib.cornell.edu/usda/AgCenusImages (Accessed 20 January 2018).
Voicu, I., and V. Been. 2008. The Effect of Community Gardens on Neighboring Property Values. *Real Estate Economics* 36: 241–83.
Watson, B. 2016. The Troubling Evolution of Corporate Greenwashing. *The Guardian*, 20 August. www.theguardian.com/sustainable-buiness/2016/aug/20/greenwashing-environmentalist-lies-companies (Accessed 9 August 2018).
Weber, R. 2002. Extracting Value from the City: Neoliberalism and Urban Redevelopment. *Antipode* 34: 519–40.
Weissman, E. 2015a. Entrepreneurial Endeavors: (Re)Producing Neoliberalization Through Urban Agriculture Youth Programming in Brooklyn, New York. *Environmental Education Research* 21: 351–64.
Weissman, E. 2015b. Brooklyn's Agrarian Questions. *Renewable Agriculture and Food Systems* 30: 92–102.
White, M. 2011. Sisters of the Soil: Urban Gardening as Resistance in Detroit. *Race/Ethnicity: Multidisciplinary Global Contexts* 5: 13–28.
Wilson, D. 1987. Urban Revitalization on the Upper West Side of Manhattan: An Urban Managerialist Assessment. *Economic Geography* 63: 35–47.
Wolch, J.R., J. Bryne, and J.P. Newell. 2014. Urban Green Space, Public Health, and Environmental Justice: The Challenge of Making Cities 'Just Green Enough'. *Landscape and Urban Planning* 125: 234–44.
Wolch, J.R., J. Wilson, and J. Fehrenbach. 2005. Parks and Park Funding in Los Angeles: An Equity-Mapping Analysis. *Urban Geography* 26: 4–35.
World Bank. 2019. *Employment in Agriculture*. Washington, DC. https://data.worldbank.org/ indicators/SLAGR.ZS?view=chart (Accessed 12 July 2019).

Yakini, M. 2013. Building a Racially Just Food Movement. *Soul Fire Farm Newsletter* #2, 26 June. www.soulfirefarm.com/newsletter-2-june-26-2013/ (Accessed 15 January 2017).

Yang, L. 1982. Community Effort Keeps a West Side Garden Flourishing. *The New York Times*, 26 August: C10.

Zezza, A., and L. Tasciotti. 2010. Urban Agriculture, Poverty, and Food Security: Empirical Evidence from a Sample of Developing Countries. *Food Policy* 35: 255–73.

Zipp, S. 2013. The Roots and Routes of Urban Renewal. *Journal of Urban History* 39: 366–91.

Zukin, S. 1987. Gentrification: Culture and Capital in the Urban Core. *Annual Review of Sociology* 13: 129–47.

4

THE CITY AS ECOSYSTEM AND ENVIRONMENT

As other ways of producing food, urban cultivation must reckon with not only relations of power in society, but also the rest of nature, or, from this point onwards, what we prefer to call biophysical reality. By biophysical we intend both ecological and physical environmental dimensions. In previous chapters, no such distinction was made because it was not analytically necessary. However, when delving into matters other than social ones we find it useful to distinguish between the worlds of organisms and those of physical environments. The term "ecological" in this chapter refers to the relations among organisms (including us) and between organisms and physical processes. Ecologies always include physical components as well, like ponds, sediments, and air temperatures. Physical processes include both forces (e.g. moving water, wind, solar radiation) and environments (e.g. rock outcrops, air spaces within soils). The difference between the two is really a function of what one asks. River water, for example, is an environment as well as a force. If one is interested in studying how fish live, then running water is an environment. If one is keener on studying how river water shapes the lay of the land, as in the formation or destruction of riverbanks, then running water is a force.

Having this in mind, one can grasp that cities are not just made of high densities of people, the variable power relations and ideas among city inhabitants, and various kinds of built environments and materials. Cities are simultaneously ecosystems and physical processes. They are composed of multiple, interacting, mutually shaping, and interdependent life forms, including people, and their physical environments (Douglas 2011a; McIntyre, Knowles-Yánez, and Hope 2000). Yet, the vast majority of those who promote urban food production treat cities as exceptional, if not separable from nature. This is evident even in the very rare works that combine social and biophysical science approaches to urban cultivation, where the two are neatly demarcated into different book chapters (e.g. Bell et al. 2016). But the meaning of ecology, over time, has broadened and specialised to include political

DOI: 10.4324/9781003131281-4

ecology, social ecology, and urban ecology. Through our analysis of urban food growing, these other ways of understanding ecology are also covered, as well as all its levels—micro, meso, and macro.

Seeing the city beyond society and polity

To some, cities are part of "second nature" on account of the preponderance of human-created surroundings (e.g. Lefebvre 1974, 368; Viljoen and Bohn 2014). The biophysical basis of cities, if it is at all recognised, is consigned to a passive background or stage over which an active, bustling humanity (often undifferentiated in mainstream discussions) can do what they please with the rest of nature. This is sometimes made obvious when it is asserted, for example, that socially oppressive systems "shape . . . environmental systems" and "environmental inequities" have social "structural roots" (Reynolds and Cohen 2016, 13–14). The latter (environmental inequities) is certainly the case, and it is crucial to underline, but the environmental effects of social structures are not as pliable as implied in such statements. This kind of reasoning has deep roots and has not spared the left. After all, the focus of leftist activism and theorising has been, from the start, issues of social inequalities and justice, and rightfully so. But those issues do not seamlessly transfer to biophysical relations. For instance, to Lefebvre, a favoured theoretician of the urban among leftists, the biophysical is either not discussed as part of cities or it is construed in metaphorical ways in the processes involved in creating urban spaces. For example, in his spider allegory for understanding body–space relations spiders are undifferentiated, their role in city ecosystems unrecognised (Lefebvre 1974, 173–5).

Lefebvre's lack of appreciation for biophysical processes (and nearly all leftist theorists on the urban) is not only repeated in current discussions of urban struggles and food production, but at times even exacerbated. Eizenberg, who draws from Lefebvre's notions, amplifies the problem by claiming that, in cities,

> Material space is an actual space of fixed, identified, and discrete entities. It is a space of experiences and practices and is therefore defined by its use-value.
> *(Eizenberg 2013, 106).*

On the same page, the author tells the reader that, when it comes to urban gardens, such a material space is composed of "soil, plants, animals, and people," all presumably "fixed, identified, and discrete." Such a view of fixity distorts and contradicts Lefebvre's understanding of spaces as continually produced socially—and therefore hardly fixed materially. This misreading of Lefebvre, however, also reinforces the view that non-human organisms and environments are passive things, analytically reducible to what is useful to and shaped by people. While urban food-growing plots are clearly human-focused, they are rooted in variable biophysical conditions (e.g. rainfall) and entities (e.g. soil nutrient levels) that bear directly upon what is grown, as well as on the health effects of its consumption (e.g. via contamination).

It is also the biophysical processes within and beyond the wider urban ecosystems that determine the fate of these plots.

Certainly, cities express extreme and concentrated forms of impact by a single (human) species, but what happens biophysically makes of that impact something that is not just willed into existence. The biophysical characteristics of places affect the very structure of a city and its everyday activities, such as what kinds of buildings can be sited where, the amount of energy needed, or how water is accessed relative to the distribution, timing, extent, and form of local water supplies. Physical processes make the social consequences of such impacts not entirely predictable or malleable. One can think of radon in basements (or other bedrock sources of pre-existing harmful materials), or earthquakes and hurricanes that lead to the sudden release of otherwise relatively contained toxic substances, or downwind effects from industrial plants that redistribute harmful chemical emissions to only some areas, or cracks in concrete or asphalt caused by rooting plants (Lundholm 2011). In urban food-growing plots, soil characteristics are of critical concern, especially as they can contain contamination from such elements as lead, arsenic, or cadmium. In such cases, contaminants in soils are derived from below-ground strata (as geogenic sources) and are not caused by humans—who determine only their exposure to them.

It is not always possible to distinguish biophysical from social causes and effects because we live in ecosystems and hence physical environments. In urban cultivation, choosing what vegetables to grow and the techniques to grow them are certainly decided by social position resulting from relations of power (e.g. Pearsall et al. 2017), much like water withdrawal and distribution rights (see Swyngedouw, Kaika, and Castro 2002). Such decisions are based as well on plant characteristics and the state of wider environmental conditions. These conditions result from biophysical sources (which may be altered by human impact), such as seasonal temperature ranges, rainfall timing, and soil characteristics.

Developing an appreciation of biophysical context helps bring about an understanding of cities as biophysical transformations with political repercussions that end up in ecologically consequential local impacts, including on human health. This, not coincidentally, is what we think reflects an ecosocialist historical and dialectical reading of the city. There is much at stake here beyond individual cities, too. In light of global warming, the impacts of what happens in cities are also ecologically global, notably their GHG emissions. This dialectical view impels us to pay attention to how we, as with other biota, and physical processes are linked and interdependent, continuously remaking a complex existing whole that is not fixed and immutable (Haila and Levins 1992; Levins and Lewontin 1985). Ecosystems are made of interconnected, interacting, and interdependent parts. They are dynamic and shifting, even if at different paces (some imperceptibly gradual, some faster than a blink of an eye), and cities are no exception. Climate change is a prime example of this dialectic at work, as it features momentous biospheric changes resulting from interactions between human societies and biophysical entities—and in turn it reshapes the conditions of life.

Given the many forces involved, there are always unintended consequences to human actions. Within a dialectical framework, political and social practices cannot be looked at as if separable from ecosystems, which is a standard (capitalist) and fundamentally flawed attitude towards not only nature but also ourselves. A view of environmental (or ecological) justice that overlooks biophysical processes and/or our inseparability from them will be unable to distinguish between human health promoting and undermining transformations of cities. An example is the evolution of human activities relatively new to some places (e.g. urban cultivation) that heighten risk of exposure to toxic substances resulting from prior human impact and/or pre-existing rock formations enriched in elements that compromise our health (e.g. high levels of nickel). Ignoring this risk undercuts political projects aiming to open up cities to more sustainable practices and their beneficial effects, including more socially inclusive and healthful habitation. This outcome will depend not only on what people do (and think) but also on much wider ecological relations in urban environments. An example of the latter is the degree of microbe persistence or proliferation, microscopic predator–prey relations, and degree of environmental conditions favourable to some microbial populations over others, all of which may be affected by human activities. A dialectical approach clarifies actual and potential ecosocial changes and consequences, e.g. as the ecological changes wrought by political projects also result in political changes. For the purposes of our task in studying urban food production prospects and objectives, a dialectical approach can help to formulate relevant technical interventions that serve to open up social and political possibilities which can be applied to everyday struggles for producing and accessing food.

In the rest of this chapter, the general biophysical characteristics of cities are discussed to provide a general background to the main ways in which biophysical processes affect urban food production. The influences go the opposite way as well. Cultivation in cities has ecologically transformative consequences, too, within as well as beyond urban centres. The ramifications of urban food production bear upon air, soil, and water quality. There can be problems with water access on account of the ways it is distributed or sourced. There can be various kinds of pollution of air and soil that impair crop growth or contaminate the food produced. Much attention here is given to trace elements (like arsenic, cadmium, lead, etc.) because they present the greatest pollution challenge. This is because they cannot be broken down and can persist as a contamination problem for centuries. There are also cases where soil properties help to neutralise some pollutants, preventing them from becoming available for plant uptake. Understanding such biophysical dynamics affecting the distribution and levels of pollutants helps to know under what conditions food-growing environments are more or less compromised and what can be done to reduce potential exposure to harmful substances.

Biophysical relations and capitalist cities

The biophysical conditions of cities differ widely and so do cities' biophysical effects. Urban settings are as subjected to climate variability as any other kind of area, as given by the angle of incidence of solar radiation (latitude), altitude, distance to large bodies of water (maritime or continental effects), and location relative to the circulation of different types of air masses, among other dynamics (Grimmond 2011). Landform, hydrology (e.g. ground and surface water supply and distribution), surface rock and sediment formations, soil types, and pre-existing biome characteristics are among the biophysical factors that affect city development and everyday urban experience. Some cities have abundant precipitation and groundwater availability with temperate forest biomes, while other cities are located in arid lands with desert biomes. Some cities receive too much rainfall because of being situated on the windward side of a mountain range, while on the other side of the same mountain range conditions can be very dry. Soils and water may contain high amounts of contaminants because of the chemical composition of underlying bedrock. Some urban dwellers must take precautions against highly venomous spiders and snakes, while others may have to contend with poison ivy, Lyme disease from ticks, or occasional black bear and coyote encounters. Urban centres in coastal areas—as a consequence of climate change, city growth, and increasing groundwater withdrawals—become more vulnerable to hurricane impact, sea-level rise, flooding, and salt-water intrusion. Places where there are periodic droughts, increasing in frequency and magnitude in relation to climate change, become more prone to devastating wildfires, especially in cases of rapid urban expansion and/or replacement of pre-existing plant species with those that produce more combustible materials. These examples of physical environments and ecological relationships can be highly impacted by people differentially (involving multiple scales, from global to local).

Urban environments, nevertheless, tend to be places of greatly concentrated human impact and lasting alteration of biophysical dynamics. The extent to which cities are transformed physical environments and ecosystems also varies considerably. Very generally, we can distinguish between cities according to whether or not and to what degree they rely on or were established through processes of industrialisation. Industrialisation processes are inextricable historically from the development of capitalism and the state-socialist projects that attempted to diverge substantively from capitalism, most of which were undermined from within as well as from massive military pressures from capitalist powers. The technological complexes and infrastructure defining most of today's cities, including its lethal repercussions, cannot be explained without looking into processes of capital accumulation.

In our present work, we focus on industrialised cities to address situations where environmental destruction tendencies prevail and are therefore pressing, and have global ramifications. That industrialised cities can be changed to become ecologically constructive ("smart", "sustainable", etc.), even when allegedly de-industrialised (as if there were no reliance on industrially produced commodities), is unconvincing. This is because of long-lasting pollution problems traceable, in many cases, to impacts at times more than a century ago, and because of the ecologically devastating reliance on global flows of energy and materials. De-industrialisation signifies the abandonment of major means of production, often leaving behind accumulations of toxic residues. These kinds of situations lay bare the intrinsically uneven development of capitalist relations, that is, increasing levels of material well-being in some places at the expense of other places. This happens at various scales, from neighbourhoods to countries. As capitalist investment hops from one place to another, seeking ever-higher returns, some cities, even in the global North, are impoverished.

The inter-related inequalities within and between as well as within cities (and countries) are major, even if not the only determinants of urban ecosystem differences. In many parts of the world (whether over- or underdeveloped), these include the long-lasting effects of settler colonialism. Such patterns tend to be characterised in settlements by largely paved commercial cores with enormous concrete structures and the radical alteration of ecosystem characteristics, such as species composition and soil properties (Ignatieva and Stewart 2009). Large economic disparities and, in many situations, horrific settler colonial legacies (genocides, slavery systems, species extirpations) make not only the availability of green spaces, but also resource consumption and exposure to toxic hazards highly unequal. Furthermore, the harmful substances resulting from the overall high environmental impact of cities tends to be concentrated most in low-income neighbourhoods (or transferred to poorer rural areas or to parts of other countries), as environmental justice activists have long been pointing out (Bullard 1990; Holifield, Chakraborty, and Walker 2018; Pulido 2000).

The cities many refer to as environmentally high-impacting—and which are now most cities in the wealthiest countries—are those that thrive on disproportionately large inputs of materials and energy from many parts of the world. These resources, once extracted and transformed, are mostly appropriated by means of everyday purchases of all sorts of consumer products. Impacts thereby extend much beyond such cities' immediate confines to reach virtually the entire planet. This is not only through insatiable and unequal resource consumption, but also through disproportionately large outputs, like GHG and other emissions and enormous amounts of waste (Grimmond 2011; Marzluff et al. 2008), waste that is often toxic and meted out on the politically marginalised. These impacts are well appreciated by ecologists, among others. Some categorise such cities "fossil-fuel subsidised" (Pickett et al. 2009, 45). Yet what appears little understood is that the net destructive impact of industrialised cities is intimately tied to their capitalist underpinnings, where endless wealth accumulation is the priority and social inequality the means.

Cities' biophysical characteristics

The outcomes of coupled yet diverse non-human and capitalist social systems generate highly variable biophysical characteristics in cities, all of which affect food production. Urban settings do differ from other kinds of environments because, as stated earlier, they feature a high concentration of human impact at the same time that they are shaped by pre-existing and co-occurring phenomena that may have little to do with human activities.

General ecological characteristics

Three principal characteristics make cities ecologically distinctive (Table 4.1). One is the high fragmentation that gives rise to separated sites for home, jobs, shops, etc. With countless embedded structures and heavy vehicular traffic, many biota find it difficult to get food and survive. Such physical layouts privilege birds, rodents, and other species that can climb or fly over obstacles or that can thrive in many different habitats. Other kinds of species, like fungi and lichens, often become rarefied, and some end up locally extinct. Another urban ecosystem characteristic is the inordinate amount of introduced or otherwise non-native species brought from afar and whose propagation is promoted by local ecological conditions. The result, typically, is the stifling if not disappearance of native species, even as total biodiversity may rise. Finally, biological community changes are largely shaped (though not necessarily caused) by human activities. Widely different impacts over short distances create communities of organisms with diverse species composition and trajectories. Cultivation in cities may contribute to habitat fragmentation if previous habitat contiguity is severed, such as with separation fences or walls that may be present even within areas dedicated to cultivation because of private plots or individualised allotments. Food production often leads to the introduction of non-native species (including those that feed on such plants or animals), creating habitats amenable for some species and throttling for others. Constant land reworking each year also induces potentially negative changes to many organisms, especially soil-dwelling ones. Ecologists usually refer to such phenomena as shortened or repeatedly interrupted successional sequences (Niemelä, Kotze, and Yli-Pelkonen 2009, 13–17).

General environmental characteristics

There are, at the same time, physical environment aspects that typify cities (Table 4.2). Elements (whether nutrients, or biologically non-essential or even toxic) are cycled much faster than in other kinds of environments, that is to say,

TABLE 4.1 General and Inter-Related Ecological Characteristics of Cities

Characteristic	*Process*	*Effects*
Structural obstacles and fragmentation	Survival-undermining impediments for species unable to overcome or cross structural obstacles	Rarefaction of slower-moving or less mobile organisms (lichen, fungi, molluscs, etc.)
Prevalence of non-native species	Stifling or disappearance of native species	Often higher biodiversity
Differentiation through frequent transformations of environments over short distances	Formation of many small biological communities	Biodiversity concentrated over small areas, more vulnerable to annihilation from extreme events

TABLE 4.2 General and Inter-Related Environmental Characteristics of Cities

Characteristic	*Process*	*Effects*
Accelerated biogeochemical cycling	Faster transfers of substances through air, water, soil, and organisms	Similar substances spread across different areas
Magnification of existing fluxes	Higher volumes and concentration of elements	Accumulation of inert and toxic materials faster than assimilation or dispersal rates
Highly modified environmental conditions	Influx of novel materials or elements and acceleration of fluxes	Higher incidence of flooding, air pollution, and the like
High energy input	Large amounts of imported sources of energy; high heat dissipation	Energy use an order of magnitude greater than other ecosystems; heat island effects, including temperature extremes and anomalies over short durations and distances

elements are moved about, back and forth, among the different environmental components, air, water, land, as well as through organisms. This is also called biogeochemical cycling. The acceleration of biogeochemical cycling happens with a magnification of existing or new fluxes and with a modification of environmental conditions. Existing fluxes can be in the form of rainfall and groundwater replenishment, which is sped up by higher rates of water withdrawal and greater evaporative transfer to the atmosphere as groundwater is depleted. A new flux is through the introduction and diffusion of motorised vehicles. Among other polluting processes, this has involved the release of small to very small and highly health-damaging dust particles that can be inhaled and even absorbed through the skin. These particles are 10 to 2.5 microns in diameter, also known respectively as PM_{10} and $PM_{2.5}$. There are even smaller particles emitted that are measurable in nanometres and that go right through bodies. Along with these variable-size particles, there is the spread of nickel and cadmium by tyre and brake abrasion, historically of lead by leaded petrol, and, more recently, platinum by way of catalytic converters. These elements, mostly toxic to us, are often embedded in dust particles that can then be absorbed by plants and soils.

In terms of altering environmental conditions, water flow, for example, is made faster by widespread impervious surfaces, so that elements dissolved in water travel faster through a city and beyond. Elements are transferred from outside (imported) and through cities, stored temporarily or for long periods in urban soils, sediments, water, organisms, and then eventually spread to areas close and far from cities. Carbon and nitrogen tend to be imported in the form of fossil fuels, fertilisers, and other sources that, when applied in cities, are rapidly released into the air (e.g. burning fossil fuels) and into water (e.g. water-dissolved nitrate seepage into groundwater). For some substances, like metals and petroleum-based materials, this accelerated process of input, throughput, storage, and output has been occurring in ever-larger quantities (Grimm et al. 2011; Thibodeux and Mackay 2011).

Energy fluxes

Cities generally differ from other environments in that energy consumption is at least one order of magnitude more per unit area per year than in other ecosystems, much beyond the amount of solar energy received annually (McIntyre, Knowles-Yánez, and Hope 2000). Per person, the figure can be much lower (particularly when considering blocks of flats) and even approach levels that match incoming solar radiation (unless one lives in a circumpolar or polar region). But area units are more appropriate for the purpose of ecosystem analysis since ecosystem level is defined by interactions among different species over a defined area. This analytical specificity is politically important. It vindicates Haila and Levins' (1992) insistence that social ends cannot be derived from the nature of ecological dynamics, which imply a scale of analysis beyond single species, societies, and individual organisms.

Energy consumption is anyway only one part of the ecological equation. Energy is never static; it is, in part, defined by constant movement. In cities, energy transfers and cycling happen through the use of overwhelmingly external sources and the excessive local dissipation of heat. Fossil fuel use largely provides the means by which typical ecosystem energy constraints are overcome. Ecosystems are powered by net primary productivity, that is, total yearly solar energy received, stored in, and available from (mostly) plant life. Overcoming these limits through fossil fuels implies undermining or destroying other ecosystems to extract energy sources. It also means rapidly ransacking the stock of solar energy accumulated and stored in fossilised organisms and accumulated over millions of years.

Heat island effects

Barring economic downturns, energy use within cities has steadily increased, as has heat dissipation, with localised climate-altering consequences. Net heat emission results from the interlocking activities of service-sector work, motorised vehicle and household appliance use, industrial production (including power plants), all of which involve energy-dissipative heating and cooling systems as well. Construction materials for such features as roads and buildings tend to absorb and release heat rapidly, rather than storing and gradually releasing heat. As a result of these processes, cities tend to be warmer on average than their surrounding areas, especially at night. In places where cold climates prevail, frost can arrive late and in mitigated form or not at all. In warmer climates, excessive heat can also lead to higher incidence of drought. It can affect entire regions when metropolitan areas become more or less contiguous, like the Ruhr District in Germany or part of the Northeastern US (from Boston to Washington, and possibly beyond). This urban heat island effect has been known since the 1820s (Parlow 2011, 34–7), yet not much has ever been done to reduce or avoid it except, for the most part, by creating new or enlarging pre-existing green areas. In fact, the problem has only expanded over time to engulf more cities and regions. What is more, heat differences can be pronounced along very small distances within cities because of the high variability of building materials, degree of surface reflectance, extent of vegetation cover, type of vegetation, and water availability and use, among other factors (Oke 2011).

Higher temperatures are linked to or accompanied by other kinds of influences. These include the cycling of substances and the modification of precipitation and wind patterns. Cycling of plant nutrients like nitrogen can be sped up because of higher temperatures and the consequently greater activity of decomposers (Pouyat and Carreiro 2003). Soil surfaces also dry out much more readily, and there are great temperature contrasts between soil surfaces and subsoils that stress organisms, including impairing seedling development (Marcotullio 2009). Rain and lightning may be enhanced downwind or in the perimeters of large cities. Aerosols (particulate matter) emitted in cities are typically very fine in diameter. They act to delay raindrop formation, often leading to higher-intensity rainfall, which can raise the magnitude of floods in cities near rivers (Douglas 2011b; Shepherd et al. 2011). Air can move with greater turbulence due to the presence of large buildings with relatively narrow canyon-like spaces in between. Wind speed and relative cloudiness also affect the magnitude of the urban heat island effect, with faster winds and greater cloudiness reducing temperatures (Oke 2011, 127–8).

Air pollution

Air quality, compared to urban heat island effects, has been much more acted on politically, even if it remains a recurring problem. Like industrial centres in the countryside, cities are ravaged by large amounts of emitted dust and pollutants and in socially very disproportionate ways (Buzzelli 2008). Aerosols from these kinds of sources, including vehicles, tend to be very small, as mentioned earlier. But in cities many more people are negatively affected at once, typically by primary pollutants like nitrogen and sulphur oxides, carbon monoxide, and volatile organic compounds, and by secondary pollutants like ozone and smog. It must be recognised as well that released chemicals are not necessarily noxious, at least not to all beings. Plants and micro-organisms benefit from the large amounts of nitrogen, sulphur, and carbon emitted, provided they are in the forms they can use, namely ammonium, carbon dioxide, methane, nitrate, and sulphate.

Urban air pollution can therefore simultaneously improve and undermine the lives of the same kinds of biota, as well as have a net damaging effect on us and many other organisms. Sources vary, and they can include forest fires, volcanic eruptions, and dust storms, depending on the biophysical context; however, most of the aerosols and pollutants are from burning fossil fuels and waste, tyre abrasion, industrial production processes, and other such activities. Dust concentrations tend to be much greater, and particles are constantly re-suspended and redistributed within cities. A significant amount of pollution in cities of the global South comes from the widespread use of open domestic fires for cooking food and other tasks. Breathing particulate matter, typically laden with pollutants (and sometimes pathogens), is a recurring health hazard causing millions of premature deaths or various debilitating diseases for millions more people (Laidlaw and Filippelli 2008; Loynachan 2016; Parlow 2011, 38–40).

Urban hydrology

Water in cities moves differently compared to other ecosystems; it travels over and between paved areas and is leaked from pipes and concrete structures. It is often re-directed from streams or lakes, as well as withdrawn from the subsoil from nearby and far-away places by creating deep wells, dammed rivers, and upstream reservoirs. It is then channelled and piped for redistribution and disposal (Douglas 2011b; Illgen 2011), albeit in a socially very unjust manner because of capitalist pressures, if not outright successes in privatising or otherwise controlling water access (Bakker 2003; Swyngedouw, Kaika, and Castro 2002). Hydrologically, industrialised cities feature their own sorts of over- and underground networks of often narrowly confined channels and basins, superimposed over or adjacent to pre-existing and shifting surface water patterns (some are legacies produced by people in antiquity).

In some cases, like Chicago, the local river flow has been reversed. The load of large structures can contribute to the formation of depressions (subsidence) where none existed before, and this also changes the course of flowing water (Eschman and Marcus 1972). Urban areas may be built upon destroyed aquatic ecosystems, such as lakes and wetlands, drained to make room for large structures, parks, roadways, and like features. Wetlands, lakes, and channels may even be introduced over time, forming an entirely different set of landforms compared to the obliterated pre-urban landscape (Ehrenfeld, Palta, and Stander 2011; Schaake 1972). Water quality, as in many areas of the countryside, can be severely undermined by an influx of untreated sewage. The reality of most cities is that often toxic compounds are emitted daily from innumerable sources, especially in highly marketised commodity-consuming areas.

Urban soils

Many of the processes recounted about organisms and their relationships, about landforms, elemental and energy flows, climate, and water all converge in phenomena called soils. Because urban cultivation is, by and large, premised on the availability and use of soils in cities, there tends to be profuse erosion of soil and sediment from the constant bustle of construction, which is behind sometimes excessive water turbidity and sediment influx into streams, lakes, and other surface waters. Streams, if not dredged, eventually become shallower. This magnifies the erosive impacts of extractive industries up- and downstream of cities, such as conventional farming, mining, and quarrying. As a result, regular rainfall that, in the past, posed little threat of flooding becomes an ever-looming hazard (Douglas 2011c; Eschman and Marcus 1972).

Because of the predominance of paved (sealed) areas in industrialised cities, water in various forms (raindrops, river overflow, etc.) tends to travel and accumulate on the surface instead of at least partly infiltrating soils or sediments. This, coupled with high amounts of groundwater withdrawals, can impair groundwater recharge and lead to permanently lower water tables, a salient problem in already dry regions. In coastal areas, these impacts can also result in seawater seeping into groundwater supplies. In cities with moderate-to-high yearly precipitation or in arid climates with occasional but intense rainfall, there tends to be a higher frequency of floods and much deeper flood-water levels than would occur if urban surfaces were more water-absorptive. Similar problems happen with compacted, clay-rich, or finer-grained (silty) soil or sediment, often caused by the constant construction related to commercial functions and land speculation. In these kinds of situations, surface openings are either too narrow or blocked entirely, so that water infiltration is impeded or too slow. There may be layers underneath the surface that slow or block the downward percolation of infiltrating water. This can be due to physical conditions before a city was built, but it can also happen when sediments are dumped on top of compacted, cemented, or otherwise less pervious materials or where soils form over decades on those kinds of materials. When it rains or when a river overflows its banks because of too much rain upstream from a city, the water may infiltrate soils or sediments and drain—but then begins to slow down or gets blocked when meeting the impervious or less porous layers below so that it accumulates until no more can enter. The net result, as in the case of paved surfaces, is a much greater occurrence of flooding with more devastating impacts (Douglas 2011b; Illgen 2011).

Soils, as in the case of other kinds of terrestrial ecosystems, affect much of the movement of water within and across cities, but they are involved in much else too. All sorts of substances, from plant nutrients to organic and inorganic contaminants, are stored and released through soils. Beyond the manifold support soils offer plants, many organisms find their main domicile within soils for at least part of their lives. Soils can be thought of as an intermingling of all biophysical processes, or as evolving relations among biota, mineral and organic materials, water, and air. The mineral part can be made of broken-down (weathered) rocks or loose or compact sediment (eventually what becomes of weathered rocks). Soils contain air that is typically heavy in carbon dioxide content, as well as water that can be attached to tiny particles for decades. What is more, there are often millions of organisms of many sizes representing thousands of species. Soils can be much more biodiverse than what lies above them. The organisms that dwell in soils are constantly exchanging energy and myriad elements, as well as breaking down and storing all sorts of substances. The organic matter that forms largely out of microbial organisms is what gives soils the ability to store a great deal of water and nutrients (as well as contaminants), to provide many tiny and unique habitats, and to show their more obvious characteristics, such as darker coloration and at times sponginess towards the surface. These different characteristics are of major importance to urban cultivation, as they affect the kinds of plants that are feasible to grow and the extent to which toxicity can be mitigated.

Since they are subjected to similar kinds of impacts, soils in industrialised cities tend to develop characteristics that differ from those of other ecosystems. As described earlier, soils may allow water to infiltrate and percolate with relative ease or to block water flow, depending on the prevailing particle size (soil texture) and even more so the degree of particle aggregation (binding), along with aggregate shape and size (soil structure). In cities, this relationship between water storage and flow and soils becomes more intricate because of the constant mixing, excavation, and dumping of materials, implying the frequent erosion, disappearance, or burial of soils, as well as the formation of new soils. The parent material (the stuff out of which soils form) may also be composed of construction debris, broken-up asphalt, landfill materials, and other human-produced substrates aiding soil formation. Soils tend to be compacted because of low levels of microbial activity and therefore less organic matter content. Particles dislodged by direct raindrop impacts, due to sparse to no protective vegetation cover, collect on soil surfaces and block the openings between soil particles. When combined with water-repellent and waxy covers (formed by deposited petroleum-based substances), displaced particles can accumulate to form impervious surface crusts. Soils also tend to be interrupted by buildings, walls, and other lasting obstacles that impede water and air flow across soils, a process that contributes to greater flooding potential. Levels of soil pH tend towards the extreme. Building materials (often carbonate-rich) and large amounts of ash, salts, or pollutants liberated by many kinds of urban activity result in often very alkaline conditions. In some cities, the opposite tendency of much lower (acid) pH has been noted, related to acidic building materials or substances emitted by local industry. These inputs and extreme pH values make it difficult for a lot of plant and soil-dwelling species to survive because of low-nutrient availability, if not toxic conditions (Marcotullio 2009).

Among the most destructive impacts on soils is their partial or total sealing (paving over), which leads to the reduction of microbial activity and undermines plants' ability to get nutrients from the soil. Partial sealing can enhance contaminant concentrations, while complete sealing contributes to flooding potential and suppresses plant life (Charzyński et al. 2018). Finally, soils in industrialised cities, even those with scant histories of manufacturing, tend to have much higher concentrations of persistent toxins, especially heavy metals, and PAHs—as can be found in areas impacted by manufacturing, transportation routes intended for motorised vehicles, and military and mining activities (Alloway 2013; Meuser 2010). There are also tens of thousands of industrially produced substances, more often organic (as in carbon-based), that enter soils, as well as water. These compounds or their degradation by-products may be toxic in themselves or in combination with other compounds, and they may remain in soils for days to hundreds of years or be leached out into groundwater (Sauerwein 2011). All this makes for much greater heterogeneity across very short distances of water (and air) flow and storage as well as biological communities within and on soils, compared to soils typically found in the countryside. This helps to explain how urban food-growing sites present such great variety in the characteristics of their soils—and why their quality must be always empirically assessed rather than assumed from large surveys. Urban soil surveys are anyway still a very rare breed.

Differences among soils in cities are not just due to, sometimes, thousands of years of intense human impact. They are also the result of other, general interrelated factors that lead to the creation and development of soils over time, namely, climate, topography, originating materials, and the activities of other organisms. So, for instance, some urban soils tend to be poorly drained because of their position at the bottom of a slope near a lake, or they are clay-rich on the surface and so restrict water infiltration. In cities constructed over areas with soils high in sulphate and falling water tables there can be precipitous falls in pH to levels that can even corrode the foundations in buildings (Engel-Di Mauro 2011). Urban soils may even be contaminated with some heavy metals because of forming out of parent material derived from local underlying rock formations filled with those heavy metals (Bourennane et al. 2010; McKone et al. 2011; Pouyat and Carreiro 2003).

Ecological processes and urban soils

As shown earlier with passing examples, physical processes combine with the activities of organisms (aside from people) to produce distinctive communities of life forms and physical environments that may or may not feature evolved fragments of pre-existing communities. Modified physical processes (e.g. higher temperatures, shifting rainfall patterns, soil property extremes) may favour introduced species, whether animal or plant, over pre-existing ones without, as already noted, necessarily reducing the overall biodiversity (Douglas 2011a; Ehrenfeld, Palta, and Stander 2011, 345–6). The course of community evolution can be altered such that it can advantage a few native plants. In places like New York City and Louisville (Kentucky), faster-growing trees can become predominant because of higher nitrate availability related to more activity by introduced earthworm species and to deposition, especially from fossil fuel combustion, whether from mainly vehicular traffic or from upwind power plants in the nearby countryside.

Insect life, however, has been noted to be greatly diminished in cities because of pollutants, habitat fragmentation, water drainage impairment (due also to compaction), and induced localised water scarcities resulting from impacts like stream diversion and wetland destruction (Carreiro et al. 2009, 324). This should be of great concern to urban gardeners, given the importance of many insects in cycling plant nutrients and the dependence of most vegetable plants on pollinators, especially when an estimated 40 per cent of insect species worldwide are threatened with extinction (Sánchez-Bayo and Wyckhuys 2019). Some insects, such as carabids, can persist in cities where patches of pre-existing forest remain (Marcotullio 2009, 178), so gardeners could contribute positively by developing and promoting agroforestry with locally evolved tree species. Many invertebrate animals and micro-organisms can nevertheless find at least some urban patches that allow them to survive and perhaps even to thrive. However, it is hard for them to migrate away when a place suddenly becomes inhospitable.

Many vertebrate species, like reptiles and amphibians, often vanish once an area is urbanised, while birds may be either slightly affected or change markedly in species composition (Clucas and Marzluff 2011; Holzer et al. 2017). On the other hand, not a few vertebrate animal species (pigeons, rats, squirrels, and some mice) can live, if not prosper in urban ecosystems. These kinds of species tend to have the following characteristics: generalist (i.e. rely on a large variety of resources and habitats); rapidly and abundantly reproductive; spread across most continents; tolerant of extreme environmental variability; and capable of relatively long-distance migration. Many of these kinds of species, however, would quickly disappear without the presence of people, as the aftermath of the 1986 Chernobyl disaster seems to suggest (Adams and Lindsey 2011).

Large-scale processes and urban ecosystems

The aforementioned highly variable localised urban characteristics are affected by much wider processes and changes. Some of these are episodic, like volcanic eruptions temporarily cooling temperatures and raising the amount of dust in the air, if not overwhelming and wiping out urban areas via lava and/or hot ash. Earthquakes, tornadoes, and hurricanes can achieve similarly destructive magnitudes and lead to drastic, if not permanent changes to city life. Adding to lethal physical environmental events, capitalist competition, within and between countries, leads to conflicts that may end up in military confrontations. These are of increasingly devastating proportions, as inordinate state investment has been continuously ploughed into developing and using technologies that are ever-more effective at killing and maiming. Among the largely premeditated atrocities committed are bombing raids and the introduction of landmines. These are among other military activities that, aside from directly murdering hundreds to thousands of people, leave short- to long-term structural damage as well as lasting life-threatening dangers from unexploded ordnances and toxic substances (Coward 2010; Hewitt 2009). In Italy alone, for example, it is estimated that at least 24,000 unexploded ordnances remain buried underground in major cities like Milan, Naples, and Rome, out of more than 1 million bombs dropped over the span of World Wars I and II (Fioravanti 2018).

Aside from destroying life, especially the lives of the least empowered (Djoudi et al. 2016; Wisner et al. 2003), such events can permanently alter, damage, or destroy other organisms' habitats, leading to changes in the local composition of species within different parts of urban ecosystems. They can also lead to such physical environmental changes as polluted water, altered river courses, de-vegetated areas prone to erosion, and heightened localised concentrations of trace elements like lead or mercury. Some catastrophic events, those related to weather, are linked to global environmental changes, which must also be considered when examining urban biophysical systems. Over the past couple of centuries, GHG emissions in general and especially from cities, as well as widespread deforestation and other major alterations to land cover, have been contributing to shifts of planetary consequence, namely by way of global climate change through increasing average global temperatures (radiative forcing).

These have differing repercussions for cities, depending on their location (Childers et al. 2015; Grimm et al. 2011; Wilby and Perry 2006). Urban heat island effects and pre-existing aridity are exacerbated, and the location, timing, duration, and intensity of precipitation and, where applicable, frost and ice cover, have been shifting. Heat dissipation is likely to increase as air conditioning and refrigeration use rises with warming trends, further magnifying existing urban heat island effects. In cities at high latitudes and elevation, where urban heat island effects already attenuate winter temperatures, frost may start later and last for shorter periods or may disappear. In tropical and subtropical areas, extreme weather events have become more frequent and greater in magnitude, leading to greater stress on many organisms through the destruction of habitat or a rapid shift in the location of food sources. Water temperature can increase further, while water flow and depth, as well as water quality, may decline seasonally or overall. These tendencies undermine prospects for fish, freshwater invertebrates, and amphibians, with cascading effects on all their predators, including people relying on them for their livelihoods and/or as dietary supplements. Bird migration patterns can be radically altered, as nesting and egg-laying and wintering possibilities become compromised in temperate-climate cities. With concomitant sea-level rise in largely tropical and subtropical regions, coastal cities experience more frequent and more erosive flooding as well as seawater incursion into freshwater supplies, which can also affect irrigation systems, extend the geographical reach of untreated sewage and, where applicable, strain if not disrupt sewage treatment plants.

Flooding is a major problem for urban living in the face of changing climates. Greater flooding frequency and magnitude, including from hurricanes and severe storms, results in dredging up nearby underwater sediment that may be highly polluted, as seen with the 2012 hurricane Sandy in New York City. Greater flooding magnifies soil erosion in areas exposed to increasing storms and higher-intensity rainfall, while soil sealing, as emphasised on several occasions already, makes such flooding even more destructive. Places experiencing decreasing precipitation or higher incidence of protracted drought may also feature more wind erosion as vegetation becomes sparser, unless efforts are made to ensure soils are covered by sufficiently protective plants year-round. Here, permaculture, for example, is decidedly helpful, depending on what kinds of plants are used.

These are but some of the salient kinds of effects of global climate change on the food-growing environments and ecosystems that make up cities. At the same time, global warming trends are largely rooted in the environmental effects of capitalist society, which, because of its basis on endless profit-making (through "economic growth"), leads to endless production of things to sell (hence the development of technologies for mass production and consumption, i.e. industrialisation), requiring insatiable energy demands (hence the reliance on transportable and high energy yielding sources per unit volume, i.e. fossil fuels). The connection between capitalism and global climate change was initiated in the 1700s, gradually developed, and sped up with mass consumption of fossil fuels, particularly since the 1950s. Industrialised forms of urbanisation are tethered closely to these ever-larger energy demands. The transformation of the chemical composition of the atmosphere (leading to global warming), brought about by capitalist relations, pushes more fossil fuel consumption and ever-greater GHG emissions in cities as more energy is used to travel about for everyday needs, to keep perishable or frozen food items or medicines refrigerated, and to protect against increasingly frequent heat waves using air conditioning.

In this manner, global warming and local urban heat island effects are promoted simultaneously, bringing changes to urban ecosystems that, as they are modified, contribute to bringing about changes to the atmosphere and to other ecosystems affected by global warming. Technical interventions like expanding green spaces, including tree cover and urban gardens, implicitly question the viability of economies based on fossil fuels, but cannot go far enough until such interventions prioritise the entire urban ecosystem, including all human inhabitants. Then it becomes, or should become, evident that creating more green spaces must account for everyone's needs, which inevitably calls up the issue of wealth redistribution, since it underlies who controls the fate of urban land. The technical understanding of urban ecosystems is necessarily politicised because people are part of ecosystems, and decisions regarding who gets what resource and how much are the prime concern. This brings us closer to the next step, which is to question the legitimacy of endless capital accumulation and capitalist relations generally, and to struggle for a sane ecological future.

The recalcitrant contaminant

The ecological relations and physical environments that characterise cities make for a sort of intricacy and dynamism arguably unparalleled by other ecosystems. However, part of what makes cities—especially industrialised cities—biophysically stand apart is not really to our collective benefit. As was alluded to earlier, cities are typically plagued by contamination and pollution, the difference being that a contaminant becomes a pollutant when they reach full-fledged (often lethal) toxicity levels to us generically. This problem of contamination and pollution is a trait that urban ecosystems partly share with ecosystems affected by mining, warfare, agrochemical-intensive farming, and other kinds of ultimately life-undermining activities. Contamination by toxic or potentially toxic substances merits particular attention because of its persistence and cumulative nature. Toxicity, however, depends on the type of species and the scale of analysis (populations of one species, entire ecosystem, etc.), so what is harmful to one kind of organism or at one scale of analysis may not be to other kinds of organisms or at other scales of analysis. It is therefore important to specify and justify what one is considering. In our case, we lay emphasis on human health and what enables it at the scales of the food-producing areas and ecosystems, also including other organisms beneficial to human health and food production. This is justified by the subject matter, urban cultivation, where people are the differentiated protagonists in a story that always unfolds with other beings having widely different needs that are, in most cases, just beginning to be understood. The story is necessarily simplified here (otherwise we would be writing an encyclopaedia), but wherever possible we try to attend to this ecological diversity.

Contaminants can be traceable to pre-existing geogenic sources (that is, from underlying rocks or sediments) and magnified by biophysical processes. For example, acid rain unrelated to human impact can facilitate some heavy metals' entry and accumulation in many organisms' bodies. However, the problem is overwhelmingly due to human impacts. Contamination often proves hazardous to our health and is greatly heightened for some communities by socially caused differences in vulnerability such as who lives and works in what sort of land in which part of town, and who has access to health or ailment-preventive care. Over the past couple of centuries most cities have become harbingers of toxicities that undercut the lives of those who have been oppressed and marginalised through capitalist relations (McClintock 2015; Pulido 2016).

Toxic or potentially toxic substances, by now, number in the tens of thousands. Since the advent of industrially synthesised chemicals in the late 1800s, the vast majority are from activities in capital-rich societies. Toxins can be divided according to whether they are inorganic and organic. Inorganics include substances that do not have carbon as part of their chemical composition. These can be single elements or compounds generally unnecessary for organisms' health or are only needed in trace amounts. Organic, in the case of contaminants, does not refer to an agrochemical-avoiding food-producing technique (like organic farming) but to whether substances are structurally formed by carbon atoms. Polyclyclic aromatic hydrocarbons (PAHs) and polychlorinated biphenyls (PCBs) would fit the bill, for example. The overwhelming majority of contaminants are organic. However, unlike inorganic toxins (often heavy metals), they can be, at least potentially, broken down to smaller compounds (especially by micro-organisms and fungi) that can be less toxic or relatively innocuous. This is why inorganic (trace) elements tend to accumulate and become highly concentrated over time. As atoms bound up with or stored in one or another bunch of atoms, they cannot be broken down without being under some extreme conditions, typically unliveable for us, such as when there is rapid radioactive decay. Once concentrated, inorganic (trace) elements cannot be diluted easily by biophysical process. In this way, trace elements become the most recalcitrant, long-lasting hazard.

At the same time, there are instances in which organic substances yield hazardous substances. Nevertheless, they can be (at least potentially) degraded into smaller, harmless parts or elements (Burgess 2016, 85). In the case of inorganic elements there is no possibility of breakdown unless they are made into even more harmful radioactive elements, decaying to become eventually less toxic atoms (taking from decades to thousands of years, depending on the element). Vegetables can be grown without agrochemicals (organically) and yet be unhealthy to us because of trace element contamination. This hazard stems from soil conditions favouring root absorption and if plants tend to be accumulator species, or from pollutants incorporated into vegetables from the air or from watering that carries contaminants.

What can make things confusing about the organic–inorganic labelling system is that carbon-based substances in soils are themselves separated according to organic and inorganic carbon components. This is done to distinguish the material that comes from organisms (e.g. soil humus) apart from that which comes from minerals (e.g. carbonates). It is important to do so to make sure one can account for the chemically unique aspects of soil organic matter. Without making this distinction, one would get a false notion of a soil's fertility and water-holding capacity. Compounds that contribute to the soil organic carbon fraction can then be separated into non-toxic and toxic or potentially toxic. In the end, organic matter is ecologically beneficial, but organic substances like PAHs and PCBs are contaminants because they are generally detrimental to organisms like us.

Organic contaminants encompass a wide variety of chemicals, mostly of synthetic origin. Most were unknown to our planet until the last century and a half, and they are traceable overwhelmingly to capitalist societies. Industrialised state-socialist societies have been comparatively lesser contributors to the problem and one, Cuba, is now arguably among the least-polluting countries (see also Engel-Di Mauro 2017). In industrial cities, generally, various industrial sources release organic contaminants, from manufacturing and coal burning to incineration and waste disposal. Vehicular emissions contribute organic pollutants, as do everyday household activities (e.g. cleaning) involving, among other products, plastics, lubricants, refrigerants, solvents, preservatives, and agrochemicals.

Contaminant sources include legacies from past land use and proximity to polluting sources (e.g. incinerators, petrol stations, power plants, vehicular traffic). World trade and travel, as well as wind, may carry such contaminants far and wide. The high number of hazardous organic compounds makes it unfeasible to study multiple, combined toxicological effects (just the total number of permutations to consider is possibly unfathomable). The health effects of soil-borne organic contaminants are even more difficult to evaluate because of confounding factors from substance-altering soil properties and organisms. This is aside from an inadequate amount of research, especially on prolonged exposure at low concentrations (Burgess 2016; MacLeod et al. 2011). Even so, and on the basis of largely single chemical compound studies, many organic contaminants have been found to cause diverse types of cancer, hormonal disruption, reproductive and immune system disorders, neuro-behavioural impairment, endocrine disruption, and birth defects. These adverse health repercussions are mediated by additional factors that are simultaneously social and physiological, such as exposure length and level, age, and genetic inheritance. Toxic effects are particularly worrisome because such compounds have been and continue to be introduced in great quantities and over time periods much shorter than most organisms' (including our own) capacities to evolve ways to neutralise them. In the case of persistent organic pollutants, those resistant to most biophysical forces, the consequences will last for decades and centuries to come.

Some of these compounds have become relatively well-known noxious legacies for posterity—DDT, PCB, dieldrin, hexachlorobenzene, and many types of dioxins (in part produced by waste incinerators).

Inorganic contaminants (or trace elements) can also form part of the ingredients of products. They include heavy metals and metalloids (e.g. arsenic) as well as radioactive substances that behave like heavy metals once in soil. In themselves, trace elements are not necessarily problematic. In fact, many of them are important micronutrients (if in small amounts, such as 1 mg or less per kg of biomass) for many organisms, including us. Copper, iron, and zinc, for example, are essential for both vascular plants and vertebrate animals. Even chromium (its trivalent form, the hexavalent variety being toxic), molybdenum, selenium, and vanadium are important to our health.

The hazard of toxicity is relative to quantity. It is when present in large amounts that trace elements become toxic and because they are single elements (rather than compounds), prospects for breakdown are infinitesimal. Sources of soil trace element contamination are similar to those of organic contaminants, but additionally they can include pre-existing substrate (geogenic) sources unrelated to human impacts (Alloway 2013; Bourennane et al. 2010; Bramwell, Pless-Mulloli, and Hartley 2008; McKone et al. 2011; Nadal, Schuhmacher, and Domingo 2004; Pouyat et al. 2007; Säumel et al. 2012). People affected by high trace element levels can suffer long-term health impairment or shortened life spans. Various forms of cancer are traceable to arsenic and nickel toxicity; neural damage occurs with arsenic, mercury, and lead contamination; and kidney damage and bone fragility result from high levels of cadmium exposure (Centeno et al. 2005; Kabata-Pendias 2011). These are among the reasons that trace elements have captured much attention in assessing the viability of urban food production (Meuser 2010; Wortman and Lovell 2013).

There are many ways people can be exposed to contaminants over time. One is simply by drinking or bathing in contaminated water. Contaminated water likely affects two out of seven people in the world. It is generally a major pathway for pathogens due to inadequate or missing sewerage treatment infrastructures, but there are also many organic and inorganic contaminants that can be found in untreated (and sometimes even in treated) water supplies. It is a scourge that goes well beyond cities (see WHO 2019) and that can also affect cities in the most capital-wealthy countries (Pulido 2016). Breathing city air (or air downwind from major industries or conventional farms in the countryside) also leads to contaminant exposure. Atmospheric sources are multiple, and both organic and inorganic contaminants can get into us by inhaling and/or ingesting the airborne dust. The dust comes in many sizes, and the most toxic particles are those that are smaller than 2.5 microns in size. The even tinier nanoparticles (1–100 nanometres in diameter), which have been increased in industrial use recently, penetrate readily through our skin and can get permanently lodged within our cells. Just as with other dust particles, nanoparticles carry both beneficial and hazardous elements (Crisponi et al. 2017; Elsaesser and Howard 2012). Dust can come into cities from long-range wind transport

(Hooda 2010, 4; McKenna-Neuman 2011) as well as from local re-suspension and emissions, which are relatively more abundant and health-threatening (Alloway 2013, 25; Brevik and Burgess 2016, 73; Laidlaw and Filippelli 2008; McLachlan 2011, 141; MacLeod et al. 2011; Wortman and Lovell 2013).

In soil, the pathways to exposure are much more complicated than through water and air. Organic contaminants can be degraded by micro-organisms, depending on optimal conditions such as ample supplies of nitrogen (typical for cities with a lot of vehicular traffic using fossil fuels). Any trace elements within the organic contaminant can, in this way, be released as well. This happens, for example, with the biodegradation of sewage sludge, where heavy metals such as cadmium can be dislodged and enter the soil environment. Otherwise, heavier (in terms of molecular weight) and non-polar (hydrophobic, lipophilic) substances tend to be more recalcitrant to breakdown or dissolution, but soil organic matter may help restrain their diffusion. This is especially so in the case of non-polar compounds, which gravitate towards and get attached to other organic materials in soils. When they get attached, they are not mobile for a time and therefore cannot affect most organisms. Lighter substances often get outgassed (or can be induced into volatilising into the atmosphere, like methane) or are more easily degraded by biochemical forces into other, smaller substances that may or may not be toxic. If they are polar compounds, something else happens. Polarity means that there are more electrons bunched to one side of the compound, giving more negative charge on that side and more positive charge on the other. Polar substances dissolve more readily into soil water and any other polar substances in liquid form, and they are more susceptible to microbial as well as free-floating enzymes' attack and decomposition.

Some inorganic compounds, however, can be broken down into individual elements, including trace elements that may be embedded within. Some of these substances are highly toxic without much alteration, like asbestos fibres. These may even become attached to edible parts of vegetables as part of lodged dust particles. Generally, mineral compounds (crystalline or amorphous) are too large to enter roots unless they are dissolved or broken down to release trace elements. Even when dissolved, trace elements may not actually transfer from soil to organisms. The specific fate of trace elements in soil is not straightforward because there are many governing soil properties involved: pH, amounts and type of clay and organic matter, mineralogy, exchange capacity, and redox (e.g. waterlogged or dry) conditions (McBride 1989; Tack 2010; Young 2013). Some trace elements become soluble in soil water (hence absorbable by plant roots) under low pH (e.g. cadmium, nickel, and lead), and others become soluble under high pH (e.g. arsenic and molybdenum). Their solubility means that they are potentially plant-available, i.e. can be absorbed by roots. Some parts of soluble trace elements can be taken up or can be further modified to insoluble forms by microbes. They may get, in part, tied up by organic matter and clays or other minerals. These are only some of the complications trace elements may undergo even when soluble, and this is why a highly contaminated soil does not necessarily mean it should not be cultivated. Unfortunately, the soil tests urban cultivators receive, if they have soils tested at all, are usually limited to total heavy metal counts. Not only are such data insufficient, they can be misleading as a result of add-on effects (for an overview, see Engel-Di Mauro 2020).

General implications of biophysical processes for urban food production

Producing food in cities therefore means contending with organisms and human-transformed environments that may or may not be conducive for safe food production. In fact, cities present differing mixes of obstacles to and amenities for practising cultivation. Urban heat island effects can strain water resources and many cultivars' ability to grow, and yet enable other vegetables as well as pathogens and pests. Simultaneously, food production brings about biophysical transformations than are usually beneficial, but that may also contribute negative effects, depending on how cultivation is carried out. There may be greater carbon storage, contaminant containment, and reduced moisture loss, but also increased pollutants and pathogens (e.g. if pesticides and manures are used).

Chronic pollution problems place additional burdens on cultivators to reduce contaminant exposure to their vegetables, trees, raised animals, and to themselves. The point is that cities are hodgepodges of interacting and mutually transforming organisms and physical forces, not just human beings. Urban dwellers live in ecosystems that are highly dynamic and change in ways that, because so many beings and forces are simultaneously involved, are not pre-determined. Developing a grasp of urban biophysical processes opens possibilities to recognise and meet the challenges of producing food in cities in ways that will not imperil people producing or consuming such food. Developing such a grasp depends on there being research and testing upon which to base site-specific practices that are most effective in aiding cultivation and protecting health.

Yet, given the heavy-handed human imprint that is the city, it cannot be forgotten that urban environmental problems faced by food growers are primarily traceable to social causes. Solutions limited to technical interventions are therefore counterproductive because they fail to address and redress social problems. There are not a few constructive alternatives available to meet the challenges specific to urban ecosystems. In our view, such alternatives should be ecosocial in outlook, that is, they must, at the same time, be ecologically and socially vigilant. A way to achieve an ecosocial understanding and develop appropriate actions and practices is, like many cultural and political ecologists have insisted for decades (Robbins 2019), to learn from existing ecologically sustainable agricultural traditions common among peasant and/or Indigenous communities that have been historically oppressed and marginalised. Another, complementary approach is to adapt agroecological methods that have been developing since the late 1920s (Francis et al. 2003, 104; Gliessman 2000) to urban settings (see Chapter 6). Such an approach—currently in its infancy—is becoming recognised as an urban agroecology (Altieri and Nicholls 2018; Tornaghi and Hoekstra 2017).

References

Adams, C.E., and K.J. Lindsey. 2011. Anthropogenic Ecosystems: The Influence of People on Urban Wildlife Populations. In, *Urban Ecology. Patterns, Processes, and Applications*, edited by J. Niemelä, 116–28. Oxford: Oxford University Press.

Alloway, B.J. 2013. Sources of Heavy Metals and Metalloids in Soils. In, *Heavy Metals in Soils. Trace Metals and Metalloids in Soils and Their Bioavailability*, edited by B.J. Alloway, 3rd ed., 11–50. London: Chapman & Hall.

Altieri, M., and C. Nicholls. 2018. Urban Agroecology: Designing Biodiverse, Productive and Resilient City Farms. *Agro Sur* 46: 49–60.

Bakker, K. 2003. The Political Ecology of Water Privatization. *Studies in Political Economy* 70: 35–48.

Bell, S., R. Fox-Kämper, N. Keshavarz, M. Benson, S. Caputo, S. Noori, and A. Voigt, eds. 2016. *Urban Allotment Gardens in Europe*. London: Routledge and Earthscan.

Bourennane, H., F. Douay, T. Sterckeman, E. Villanneau, H. Ciesielski, D. King, and D. Baize. 2010. Mapping of Anthropogenic Trace Elements Inputs in Agricultural Topsoil from Northern France Using Enrichment Factors. *Geoderma* 157: 165–74.

Bramwell, L., T. Pless-Mulloli, and P. Hartley. 2008. Health Risk Assessment of Urban Agriculture Sites Using Vegetable Uptake and Bioaccessibility Data: An Overview of 28 Sites with a Combined Area of 48 Hectares. *Epidemiology* 19: S150.

Brevik, E.C., and L.C. Burgess. 2016. *Soils and Human Health*. Boca Raton, FL: CRC Press.

Bullard, R.D. 1990. *Dumping in Dixie: Race, Class, and Environmental Quality*. Boulder, CO: Westview.

Burgess, L.C. 2016. Organic Pollutants in Soil. In, *Soils and Human Health*, edited by E.C. Brevik, and L.C. Burgess, 83–106. Boca Raton, FL: CRC Press.

Buzzelli, M. 2008. A Political Ecology of Scale in Urban Air Pollution Monitoring. *Transactions of the Institute of British Geographers* 33: 502–17.

Carreiro, M.M., R.V. Pouyat, C.E. Tripler, and W.-X. Zhu. 2009. Carbon and Nitrogen Cycling in Soils of Remnant Forests along Urban-Rural Gradients: Case Studies in the New York Metropolitan Area and Louisville, Kentucky. In, *Ecology of Cities and Towns: A Comparative Approach*, edited by M.J. McDonnell, A.K. Hahs, and J.H. Breuste, 308–28. Cambridge: Cambridge University Press.

Centeno, J.A., F.G. Mullick, K.G. Ishak, T.J. Franks, A.P. Burke, M.N. Koss, D.P. Perl, P.B. Tchounwou, and J.P. Pestaner. 2005. Environmental Pathology. In, *Essentials of Medical Geology: Impacts of the Natural Environment on Public Health*, edited by O. Selenius, B. Alloway, J.A. Centeno, R.B. Finkelman, R. Fuge, U. Lindh, and P. Smedley, 563–94. Amsterdam: Elsevier.

Charzyński, P., P. Hulisz, A. Piotrowska-Długosz, D. Kamiński, and A. Plak. 2018. Sealing Effects on Properties of Urban Soils. In, *Urban Soils*, edited by R. Lal, and B.A. Stewart, 155–74. Boca Raton, FL: CRC Press.

Childers, D.L., M.L. Cadenasso, J.M. Grove, V. Marshall, B. McGrath, and S.T.A. Pickett. 2015. An Ecology *for* Cities: A Transformational Nexus of Design and Ecology to Advance Climate Change Resilience and Urban Sustainability. *Sustainability* 7: 3774–91.

Clucas, B., and J.M. Marzluff. 2011. Coupled Relationships Between Humans and Other Organisms in Urban Areas. In, *Urban Ecology. Patterns, Processes, and Applications*, edited by J. Niemelä, 135–47. Oxford: Oxford University Press.

Coward, M. 2010. Urbicide: The Politics of Urban Destruction. *Global Discourse* 1(2): 186–9.

Crisponi, G., V.M. Nurchi, J. Lachowicz, M. Peana, S. Medici, and M.A. Zoroddu. 2017. Toxicity of Nanoparticles: Etiology and Mechanisms, in Antimicrobial Nanoarchitectonics.

In, *Antimicrobial Nanoarchitectonics: From Synthesis to Applications*, edited by A.M. Grumezescu, 511–46. Amsterdam: Elsevier.

Djoudi, H., B. Locatelli, C. Vaast, K. Asher, M. Brockhaus, and B.B. Sijapati. 2016. Beyond Dichotomies: Gender and Intersecting Inequalities in Climate Change Studies. *Ambio* 45(Suppl. 3): S248–62.

Douglas, I. 2011a. The Analysis of Cities as Ecosystems. In, *The Routledge Handbook of Urban Ecology*, edited by I. Douglas, D. Goode, M. Houck, and R. Wang, 17–25. London: Routledge.

Douglas, I. 2011b. Urban Hydrology. In, *The Routledge Handbook of Urban Ecology*, edited by I. Douglas, D. Goode, M. Houck, and R. Wang, 148–58. London: Routledge.

Douglas, I. 2011c. Urban Geomorphology. In, *The Routledge Handbook of Urban Ecology*, edited by I. Douglas, D. Goode, M. Houck, and R. Wang, 159–63. London: Routledge.

Ehrenfeld, J.G., M. Palta, and E. Stander. 2011. Wetlands in Urban Environments. In, *The Routledge Handbook of Urban Ecology*, edited by I. Douglas, D. Goode, M. Houck, and R. Wang, 338–51. London: Routledge.

Eizenberg, E. 2013. *From the Ground up. Community Gardens in New York City and the Politics of Spatial Transformation*. Burlington: Ashgate.

Elsaesser, A., and C.V. HoWard. 2012. Toxicology of Nanoparticles. *Advanced Drug Delivery Reviews* 64(2): 129–37.

Engel-Di Mauro, S. 2011. Minding History and World-Scale Dynamics in Hazards Research: The Making of Hazardous Soils in The Gambia and Hungary. *Journal of Risk Research* 15(10): 1319–33.

Engel-Di Mauro, S. 2017. The Enduring Relevance of State-Socialism. *Capitalism Nature Socialism* 27(4): 1–15.

Engel-Di Mauro, S. 2020. The Troubling and Troublesome Worlds of Urban Soil Trace Element Contamination Baselines. *Environment and Planning E: Nature and Space* 3(1): 95–113.

Eschman, D.F., and M.G. Marcus. 1972. The Geologic and Topographic Setting of Cities. In, *Urbanization and the Environment. The Physical Geography of the City*, edited by T Detwyler, and M.G. Marcus, 27–50. Belmont: Duxbury Press.

Fioravanti, A. 2018. L'Italia Dorme su un Letto di Bombe Inesplose. [Italy Sleeps on a Bed of Unexploded Bombs] *La Stampa*, 28 July. www.lastampa.it/2018/07/28/italia/ litalia-dorme-su-un-letto-di-bombe-inesplose-BteVoV8MrTcer3VbFJQ45O/ premium.html (Accessed 29 July 2018).

Francis, C., G. Lieblein, S. Gliessman, T.A. Breland, N. Creamer, R. Harwood, L. Salomonsson, J. Helenius, D. Rickerl, R. Salvador, M. Wiedenhoeft, S. Simmons, P. Allen, M. Altieri, C. Flora, and R. Poincelot. 2003. Agroecology: The Ecology of Food Systems. *Journal of Sustainable Agriculture* 22(3): 99–118.

Gliessman, S.R. 2000. *Agroecology: Ecological Processes in Sustainable Agriculture*. Boca Raton, FL: Lewis Publishers.

Grimm, N.B., R.L. Hale, E.M. Cook, and D.M. Iwaniec. 2011. Urban Biogeochemical Flux Analysis. In, *The Routledge Handbook of Urban Ecology*, edited by I. Douglas, D. Goode, M. Houck, and R. Wang, 503–20. London: Routledge.

Grimmond, C.S.B. 2011. Climate of Cities. In, *The Routledge Handbook of Urban Ecology*, edited by I. Douglas, D. Goode, M. Houck, and R. Wang, 103–19. London: Routledge.

Haila, Y., and R. Levins. 1992. *Humanity and Nature: Ecology, Science, and Society*. London: Pluto Press.

Hewitt, K. 2009. Proving Grounds of Urbicide: Civil and Urban Perspectives on the Bombing of Capital Cities. *ACME: An International Journal for Critical Geographies* 8(2): 340–75.

Holifield, R., J. Chakraborty, and G. Walker. 2018. Introduction. The Worlds of Environmental Justice. In, *The Routledge Handbook of Environmental Justice*, edited by R. Holifield, J. Chakraborty, and G. Walker, 1–11. London: Routledge.

Holzer, K.A., R.P. Bayers, T.T. Nguyen, and S.P. Lawler. 2017. Habitat Value of Cities and Rice Paddies for Amphibians in Rapidly Urbanizing Vietnam. *Journal of Urban Ecology* 3(1). https://doi.org/10.1093/jue/juw007 (Accessed 21 July 2017).

Hooda, P., ed. 2010. *Trace Elements in Soils*. Oxford: Blackwell.

Ignatieva, M.A., and G.H. Stewart. 2009. Homogeneity of Urban Biotopes and Similarity of Landscape Design Language in Former Colonial Cities. In, *Ecology of Cities and Towns: A Comparative Approach*, edited by M.J. McDonnell, A.K. Hahs, and J.H. Breuste, 399–421. Cambridge: Cambridge University Press.

Illgen, M. 2011. Hydrology of Urban Environments. In, *Urban Ecology. Patterns, Processes, and Applications*, edited by J. Niemelä, 59–70. Oxford: Oxford University Press.

Kabata-Pendias, A. 2011. *Trace Elements in Soils and Plants*, 4th ed. Boca Raton, FL: CRC Press.

Laidlaw, M.A.S., and G.M. Filippelli. 2008. Resuspension of Urban Soils as a Persistent Source of Lead Poisoning in Children: A Review and New Directions. *Applied Geochemistry* 23: 2021–39.

Lefebvre, Henri. 1974 [1991]. *The Production of Space*. Translated by D. Nicholson-Smith. Oxford: Blackwell.

Levins, R., and R. Lewontin. 1985. *The Dialectical Biologist*. Cambridge: Harvard University Press.

Loynachan, T.E. 2016. Human Disease from Introduced and Resident Soilborne Pathogens. In, *Soils and Human Health*, edited by E.C. Brevik, and L.C. Burgess, 107–36. Boca Raton, FL: CRC Press.

Lundholm, J. 2011. Vegetation of Urban Hard Surfaces. In, *Urban Ecology. Patterns, Processes, and Applications*, edited by J. Niemelä, 93–102. Oxford: Oxford University Press.

MacLeod, M., M. Scheringer, C. Gotz, K. Hungerbuhler, C.I. Davidson, and T.M. Holsen. 2011. Deposition from the Atmosphere to Water and Soils with Aerosol Particles and Precipitation. In, *Handbook of Chemical Mass Transport in the Environment*, edited by L.J. Thibodeux, and D. Mackay, 103–35. Boca Raton, FL: CRC Press.

Marcotullio, P.J. 2009. Urban Soils. In, *The Routledge Handbook of Urban Ecology*, edited by I. Douglas, D. Goode, M. Houck, and R. Wang, 164–86. London: Routledge.

Marzluff, J.M., M. Alberti, G. Bradley, and U. Simon. 2008. *Urban Ecology: An International Perspective on the Interaction Between Humans and Nature*. Berlin: Springer.

McBride, M.B. 1989. Reactions Controlling Heavy Metal Solubility in Soils. *Advances in Soil Science* 10: 1–56.

McClintock, N. 2015. A Critical Physical Geography of Urban Soil Contamination. *Geoforum* 65: 69–85.

McIntyre, N.E., K. Knowles-Yánez, and D. Hope. 2000. Urban Ecology as an Interdisciplinary Field: Differences in the Use of 'Urban' Between the Social and Natural Sciences. *Urban Ecosystems* 4: 5–24.

McKenna-Neuman, C.L. 2011. Dust Resuspension and Chemical Mass Transport from Soil to the Atmosphere. In, *Handbook of Chemical Mass Transport in the Environment*, edited by L.J. Thibodeux, and D. Mackay, 453–93. Boca Raton, FL: CRC Press.

McKone, T.E., S.L. Bartelt-Hunt, M.S. Olson, and F.D. Tillman. 2011. Mass Transfer with in Surface Soils. In, *Handbook of Chemical Mass Transport in the Environment*, edited by L.J. Thibodeux, and D. Mackay, 159–211. Boca Raton, FL: CRC Press.

McLachlan, M.S. 2011. Mass Transfer between the Atmosphere and the Plant Canopy Systems. In, *Handbook of Chemical Mass Transport in the Environment*, edited by L.J. Thibodeux, and D. Mackay, 137–58. Boca Raton, FL: CRC Press.

Meuser, H. 2010. *Contaminated Urban Soils*. Dordrecht: Springer.

Nadal, M., M. Schuhmacher, and J.L. Domingo. 2004. Metal Pollution of Soils and Vegetation in an Area with Petrochemical Industry. *The Science of the Total Environment* 321: 59–69.

Niemelä, J., D.J. Kotze, and V. Yli-Pelkonen. 2009. Comparative Urban Ecology: Challenges and Possibilities. In, *Ecology of Cities and Towns: A Comparative Approach*, edited by M.J. McDonnell, A.K. Hahs, and J.H. Breuste, 9–24. Cambridge: Cambridge University Press.

Oke, T.R. 2011. Urban Heat Islands. In, *The Routledge Handbook of Urban Ecology*, edited by I. Douglas, D. Goode, M. Houck, and R. Wang, 120–31. London: Routledge.

Parlow, E. 2011. Urban Climate. In, *Urban Ecology. Patterns, Processes, and Applications*, edited by J. Niemelä, 31–44. Oxford: Oxford University Press.

Pearsall, H., S. Gachuz, M. Rodriguez Sosa, B. Schmook, H. van der Wal, and M.A. Gracia. 2017. Urban Community Garden Agrodiversity and Cultural Identity in Philadelphia, Pennsylvania, U.S. *Geographical Review* 107(3): 476–95.

Pickett, S.T.A., M.L. Cadenasso, M.J. McDonnell, and W.R. Burch, Jr. 2009. Frameworks for Urban Ecosystem Studies: Gradients, Patch Dynamics and the Human Ecosystem in the New York Metropolitan Ara and Baltimore, USA. In, *Ecology of Cities and Towns: A Comparative Approach*, edited by M.J. McDonnell, A.K. Hahs, and J.H. Breuste, 25–50. Cambridge: Cambridge University Press.

Pouyat, R.V., and M.M. Carreiro. 2003. Controls on Mass Loss and Nitrogen Dynamics of Oak Leaf Litter Along an Urban-Rural Land-Use Gradient. *Oecologia* 135: 288–98.

Pouyat, R.V., I.D. Yesilonis, J. Russell-Anelli, and N.K. Neerchal. 2007. Soil Chemical and Physical Properties That Differentiate Urban Land-Use and Cover Types. *Soil Science Society of America Journal* 71: 1010–19.

Pulido, L. 2000. Rethinking Environmental Racism: White Privilege and Urban Development in Southern California. *Annals of the Association of American Geographers* 90(1): 12–40.

Pulido, L. 2016. Flint, Environmental Racism, and Racial Capitalism. *Capitalism Nature Socialism* 27(3): 1–16.

Reynolds, K., and N. Cohen. 2016. *Beyond the Kale. Urban Agriculture and Social Justice Activism in New York City*. Athens: University of Georgia Press.

Robbins, P. 2019. *Political Ecology: A Critical Introduction*, 3rd ed. New York: Wiley.

Sánchez-Bayo, Francisco, and Kris A.G. Wyckhuys. 2019. Worldwide Decline of the Entomofauna: A Review of Its Drivers. *Biological Conservation* 232: 8–27.

Sauerwein, M. 2011. Urban Soils—Characterization, Pollution, and Relevance in Urban Ecosystems. In, *Urban Ecology. Patterns, Processes, and Applications*, edited by J. Niemelä, 45–58. Oxford: Oxford University Press.

Säumel, I., I. Kotsyuk, M. Hölscher, C. Lenkereit, F. Weber, and I. Kowarik. 2012. How Healthy Is Urban Horticulture in High Traffic Areas? Trace Metal Concentrations in Vegetable Crops from Plantings Within Inner City Neighbourhoods in Berlin, Germany. *Environmental Pollution* 165: 124–32.

Schaake Jr., J.C. 1972. Water and the City. In, *Urbanization and the Environment. The Physical Geography of the City*, edited by T. Detwyler, and M.G. Marcus, 97–134. Belmont: Duxbury Press.

Shepherd, J.M., J.A. Stallins, M.L. Jin, and T.L. Mote. 2011. Urban Effects on Precipitation and Associated Convective Processes. In, *The Routledge Handbook of Urban Ecology*, edited by I. Douglas, D. Goode, M. Houck, and R. Wang, 132–47. London: Routledge.

Swyngedouw, E., M. Kaika, and E. Castro. 2002. Urban Water: A Political-Ecology Perspective. *Built Environment* 28(2): 124–37.

Tack, F.M.G. 2010. Trace Elements: General Soil Chemistry, Principles and Processes. In, *Trace Elements in Soils*, edited by P. Hooda, 9–38. Oxford: Blackwell.

Thibodeux, L.J., and D. Mackay, eds. 2011. *Handbook of Chemical Mass Transport in the Environment*. Boca Raton, FL: CRC Press.

Tornaghi, C., and F. Hoekstra. 2017. Editorial. *Urban Agriculture Magazine* 33: 3–4. https://ruaf.org/assets/2019/11/Urban-Agriculture-Magazine-no.-33-Urban-Agroecology.pdf (Accessed 12 July 2020).

Viljoen, A., and K. Bohn, eds. 2014. *Second Nature Urban Agriculture: Designing Productive Cities*. New York: Routledge.

WHO. 2019. Drinking Water. www.who.int/news-room/fact-sheets/detail/drinking-water/ (Accessed 3 August 2020).

Wilby, R.L., and G.L.W. Perry. 2006. Climate Change, Biodiversity and the Urban Environment: A Critical Review Based on London, UK. *Progress in Physical Geography* 30(1): 73–98.

Wisner, B., P. Blaikie, T. Cannon, and I. Davis. 2003. *At Risk. Natural Hazards, People's Vulnerability, and Disasters*, 2nd ed. London: Routledge. www.preventionweb.net/files/670_72351.pdf (Accessed 17 July 2018).

Wortman, S.E., and S.T. Lovell. 2013. Environmental Challenges Threatening the Growth of Urban Agriculture in the United States. *Journal of Environmental Quality* 42(5): 1283–94.

Young, S.D. 2013. Chemistry of Heavy Metals and Metalloids in Soils. In, *Heavy Metals in Soils. Trace Metals and Metalloids in Soils and Their Bioavailability*, edited by B.J. Alloway, 3rd ed., 51–95. London: Chapman & Hall.

5

URBAN FOOD GROWING AND SOCIAL SUSTAINABILITY

While urban food growing has surged in industrialised countries—as a subject for social theory and as a local practice—inadequate attention has been given to rigorous examination of the volume and sustainability of its production, as well as of its political contradictions. A neglected subject is the changing character of urban food production around the world (Orsini et al. 2013). While it is expanding in the global North, it is stalling in the global South. Eighty per cent of the world's urban croplands are in the global South, in low- to middle-income countries (Thebo et al. 2014). Contrarily, production is declining while there is increasing use of money to buy food rather than trading for it (De Bon, Parrot, and Moustier 2010). There are multiple factors causing this trend, including land commodification, intensification of displacement of people from the countryside (often violent dispossession in order to build massive infrastructural projects or to benefit mining and forestry), residual warfare hazards (e.g. landmines), and increasing unemployment because of government policies to promote mechanised, plantation-scale commercial farming (much production of which goes to the global North). Thus, the prospects for food self-sufficiency in global South cities are declining as more and more people are forced to migrate from the countryside. This is occurring while more food growing land is converted to housing. Still, food self-sufficiency remains comparatively a far less tenable prospect in cities of the global North (Mok et al. 2014).

The food access problems confronting people in the global South remain paramount in scale and intensity, but our attention now turns to urban agriculture in the global North. Much has been made of its potentials and this necessitates closer inspection. Another reason is that global North urban agriculture will likely pave the way for the future of urban farming elsewhere, as policies devised within the wealthiest countries are often imposed on the rest of the world. Moreover, the intense land and real estate speculation and financialisation, typical of global North

DOI: 10.4324/9781003131281-5

cities, illustrate the limits of and challenges to urban agriculture in capitalist contexts anywhere.

From capitalist agriculture to ecosocialist cultivation in cities

This chapter begins with a critique of the faulty notions that underlie the optimistic enthusiasm about urban food output and ecological sustainability potentials. The optimism is evidenced by the adoption of the term agriculture to replace that of gardening as the designation of urban food growing. An example of this in urban planning discourse is that an *agricultural urbanism* has been proposed as the next big social movement in the development of cities (de la Salle and Holland 2010). Analyses of urban food growing in the global North are going through a transition, prompted in large measure by growing scepticism about its assumed potentials. There have been three broad developments in its history, stretching back to the early twentieth century. The first stage consisted of low-level growing, punctuated by sharp spurts of expansion set off by economic depressions and world wars, followed by contractions when these events ended. The second stage, beginning in the latter decades of the twentieth century, was an upsurge in gardening that was not prompted by a major depression or war. This surge has been spotlighted by a bandwagon of supporters who assume a large food production capacity as well as major contributions to ecological sustainability.

Presently, we are at the cusp of another stage shift, as a growing body of research supports a less optimistic view of urban agriculture's potentials (Sonnino 2013; Martin, Clift, and Christie 2016; WinklerPrins 2017). The shape and substance of this new, emergent realism is being fashioned by constructive critiques—and many are cited in this text. They support a similar broad conclusion and it is, in a few words, that urban agriculture overstates its food output and improvements in ecological sustainability while neglecting its social and political implications. Some critics add a further reason for a re-framing of urban food growing—its ambivalence with regard to prevailing neoliberal inequalities and injustices in food production and consumption. Our work expands and deepens the critique and moves on to envisioning a progressive ecosocial framework for urban food growing.

Our critical analysis begins with assessments of fuzzy claims for urban agriculture rooted in assumptions about food security (conventionally confused with food access) and ecological sustainability. Our evaluation is that they do not pass empirical muster. Additionally, as we point out, the views of gardeners themselves do not support the optimistic claims. Urban agriculture's overreach for output overshadows possibilities for giving direct attention to social sustainability prospects. We make a case for re-designating urban food growing from urban agriculture to urban cultivation, informed by others who have done so (such as Martin, Clift, and Christie 2016; WinklerPrins 2017). However, none have thus far focused specifically on the problems inherent to calling it agriculture. We make an extended case for cultivation based on two principal considerations: (1) urban food growing cannot reach the level of agriculture in terms of output; (2) the term cultivation puts the social strengths of urban food growing in the spotlight, ahead of output. These social strengths, such as providing for community organisation of public spaces, provide opportunities for developing our ecosocialist perspective. Part of our argument is that a singular focus on output (and on trying to raise it through new technologies) leads to the development of a profit-based urban food growing, already in progress, and obfuscates possibilities for challenging capitalism.

Land availability and appropriate growing conditions

In the global North, urban food growing has been freighted with the promise of substantially expanding the level of food production. However, this claim cannot be made even for global South cities, where urban land availability for food production is much higher—and yet cannot come close to reaching food security levels (Jácome-Pólit et al. 2019, 287; Karg et al. 2016; Wegerif and Wiskerke 2017).

The universality of local variability in urban food growing makes case studies a requirement in its output and ecological assessment (Corcoran and Cavino 2018, 4). Its sites differ operationally in their auspices, organisational format, size, and kind of urban location, as well as biophysical conditions. A major structural variable is the status of local economies, largely a product of capital flows. The general cleavage is between prosperous and distressed cities. The latter, highlighted by the archetypal case of Detroit (Mogk, Wiatkowski, and Weindorf 2010), have large dollops of abandoned land that may serve as a base for upscaling food growing (Harris 2010). The three cases used to ground the analysis here are from prosperous cities—London, New York City, and San Francisco (Martin, Clift, and Christie 2016). All three have a relatively small portion of abandoned land, and the amount is steadily shrinking. It is in such prosperous cities of the global North, where land use is subjected to the highest rent-seeking pressures, that we witness the weight of urban food growing's promotion and visibility, and of its heralded potentials.

Archival and observational data from London, New York, and San Francisco sites demonstrate that urban gardens produce only a very small proportion of just the fruit and vegetable portion of plates for residents of their catchment areas (Martin, Clift, and Christie 2016), and fruit and vegetables comprise just one-third of a healthy and sustainable diet (Macdiarmid et al. 2012). The three sites differed in their constitutions: a very small (0.05 ha) inner city community garden in New York's Manhattan Borough (Figure 5.1), a small (1.42 ha) inner suburban farm in London's Sutton Borough (Figure 5.2), and a large (6.48 ha) exurban farm within San Francisco's metropolitan area (Figure 5.3). Catchment areas are the local precincts of the sites' residents as numerated by geographical and political units, such as census tracts.

The data analysis was based on metrics used by Garnett (2000) and WHO (2000). The major finding was that it would require over 4,500 sites to provide fruit and vegetables for residents of the catchment area of the inner-city community garden; over 400 sites for that of the suburban farm; and over 1,000 sites for that

FIGURE 5.1 Gathering at West Side Community Garden, Manhattan, New York City, US (photo by George Martin 2012)

of the exurban farm. None of the catchment areas had other, similar food-growing sites open to public use. There were an unknown number of domestic garden plots (and allotments in the London area), but it is unlikely that their output could significantly reduce the magnitudes of the estimates.

Studies of food output from other cities show varying but comparable results: Cleveland (Grewal and Grewal 2012); Detroit (Colasanti, Litjens, and Hamm 2010); London (Garnett 2000); Oakland (McClintock, Cooper, and Khandeshi 2013); and Oxford (FCRN 2012). One study found an impressive 4,600 food production sites totalling about 26.3 ha (64.9 acres) in Chicago, the bulk of which were domestic garden plots (Taylor and Lovell 2012). Martin, Clift, and Christie (2016) estimated all the sites to be capable of producing fruit and vegetables for a miniscule proportion of Chicago's 2.7 million residents. It should be noted that all estimates were based on uniform distributions across urban populations, and if produce is consumed by specific subgroups, the benefits may be more significant (see below). A Philadelphia study found that its 15.4 ha of food growing produced more than 900 tonnes (2 million lb) of food in 2008 (Vitiello and Nairn 2009). However, according to our back-of-the-envelope calculation, this would have provided only the equivalent of a daily nibble of food (about 600 g, or 1.3 lb) for the city's 1.5 million residents. The optimistic projections of a very large urban

FIGURE 5.2 Sutton Borough Community Farm, London, UK (photo by George Martin 2014)

FIGURE 5.3 Vegetables and flowers, West Side Community Garden (photo by George Martin 2014)

horticulture potential are based in studies using GIS data (Edmondson et al. 2020). However, such estimates do not incorporate field analyses of the soils under green cover.

If urban produce is grown and consumed by disadvantaged groups benefits may be more significant. One study shows that it can make a substantial contribution to the tables of low-income immigrants from agricultural backgrounds (Mares and Pena 2010). Also, the CDC (2010) in the US inventoried examples of community gardens' efforts that make substantial inputs to improving the diets of low-income persons with high rates of obesity and diabetes and with limited sources of fresh produce. However, these are special cases, in which volunteer and experienced gardeners had convenient access to free plots of arable land, not a common situation for low-income residents of cities in the global North. There are several socioeconomic factors that constrain urban food production, including a lack of skilled gardening labour and operational capital. For example, a case study of food-growing sites in Houston revealed that their major challenge—by a large margin—was finding committed and craft-knowledgeable volunteers for the substantial physical labour required to grow food (Broadstone and Brannstrom 2017). Additionally, securing money for gardening tools and supplies was a challenge for the 31 sites; nearly half relied heavily on financial support from charitable and non-profit organisations.

The fundamental limitation to urban cultivation's production potential remains a material one—the scarcity of arable land. Field production is simply beyond the land capacities even of enormous urban agglomerations, including those in the global South, where peri-urban farming is being overwhelmed by urban expansion. In the face of the optimistic scaling scenarios rooted in localism, an optimisation modelling of the production of temperate and tropical cereals, and tropical roots and pulses, demonstrated that only 11–28 per cent of them could be grown within a 100 km radius of urban centres (Kinnunen et al. 2020). Cities, by virtue of the reality of being cities, cannot grow what has become the basic food of humanity—cereal grains, the stuff of the traditional "staff of life." Cereals supply over half the dietary energy of humanity (FAO 2010), and large rural fields are the venues for their efficient and sustainable production. For example, research indicates that peak efficiency in growing is achieved with production units of 160 to 325 ha for soybean and 325 to 490 ha for maize (Duffy 2009). Such large-scale production efficiencies are critical to providing for food output, which is greatly hampered by the expansion of meat-based diets and biofuel production, and newly exposed to the extreme negative impacts of global warming (IPCC 2014; USDA 2013).

In reality, just maintaining presently low outputs of urban food production will require overcoming some daunting obstacles. Prosperous cities struggle to hold on to current growing spaces, much less developing new ones. For example, in London, the area of domestic gardens, which comprises 25 per cent of the land upon which fruit and vegetables could be grown, is declining. Between 1998 and 2008 growing space fell by 12 per cent while the area of hard surfacing increased by 26 per cent, largely paving for car parking (Smith 2010). This change corresponds to sharp rises in house prices in London over the same period. Real prices indexed for 2003 grew by a multiple of 2.4 in the same decade (Marsden 2015). Moreover, the UK's urban food-growing allotments, despite their strong historical and cultural roots, are increasingly contested spaces (Perez-Vazquez, Anderson, and Rogers 2005). They are being seen more and more as opportunities by hard-pressed local governing councils to meet housing needs (Scott et al. 2017).

Globally, the land area that would be needed to produce just the vegetables in a healthy sustainable diet would be equivalent to *about three-fourths of all urbanised land* (Martellozzo et al. 2015). Compounding this scale barrier, there is an increasing competition for land in cities around the world as they become the targets of international financiers and land speculators as more people migrate to cities. Research has consistently supported what was found for Oakland—that "even massive agglomerations of urban gardens are unlikely to meet more than 5 per cent of the vegetable demands" (McClintock and Cooper 2009; in McClintock 2011, 113). At the end of the day, the enduring universal characteristic of cities is that they are places where large numbers of people live in small areas. They do not contain expanses of ground-level land fully exposed to the sun, which are needed for the field crops required to feed their residents. Hence, urban food growing will make only a limited contribution to food security (much less food access) even in countries of the global South, China now being a prominent example (Badami and Ramankutty 2015).

The increased hard surfacing of land for motor vehicles in global North cities results in biophysically deleterious and long-lasting consequences, as does converting abandoned brownfield sites for food growing (see last chapter). One, London's Spitalfield City Farm, comprising 0.53 ha (1.3 acres), is described as a "community place" that features gardens and farm animals. It started in 1978 on land formerly occupied by a railway goods depot (see www.spitalfieldcityfarm.org). What, then, of the possibilities of creating more urban food-growing land by utilising the abandoned brownfield sites that exist in all cities of the global North? Unfortunately, their use is questionable. Brownfield soils frequently have a threatening public health legacy of toxicity, or an agronomic one of low quality for growing flora. For example, a study of lead contamination in Oakland sites found a high level of variability that must be considered with care before undertaking food growing (McClintock, Cooper, and Khandeshi 2013). Another study, in New York City, found that 97 per cent of community and backyard gardens tested had elevated levels of topsoil lead and arsenic, especially in lower-income census tracts (Cheng et al. 2015; Li et al. 2017). To underline something from the previous chapter, high amounts of soil lead do not mean that there cannot be any food grown in such areas. The lead is unlikely to be absorbed by roots unless pH is very low (very acidic soil), usually to the point of hindering plant growth anyway. In such circumstances of high lead or heavy metal contamination, root vegetables are best avoided, dust must be kept down by keeping soil always vegetated or otherwise covered (with mulch, for instance), vegetables must always be thoroughly washed (and roots discarded) before eating them, and protective gear should be worn when working the soil to prevent particle inhalation (like masks, gloves, and goggles as well as complete body covering). Nevertheless, even with all safety precautions in place, not all brownfield sites would qualify to pass a sunshine availability test.

Low land availability, toxic soils, and shaded areas mean that there are many still unaddressed obstacles limiting potential to scale-up urban food growing. Rural, large-scale farming will continue to be required to advance sustainable food access for the world's population. More equitable (Rosin, Stock, and Campbell 2012) and less wasteful (Kummu et al. 2012) food distribution is especially needed. The FAO (2017) has estimated that 815 million people (11 per cent of the world's population) were chronically undernourished in 2017. Without serious policies to ensure everyone worldwide has ready food access and without public (or, better, local and democratised) control over food systems, there is little chance of countering hunger, which is an inherent and chronic feature of capitalist food systems. The widely recognised and criticised ecological and social problems posed by massive farms are not generated by scale (InterAcademy Partnership 2018). Instead, they stem from capital-driven conditions that feature soil and plant chemicalisation, labour exploitation, and extreme food maldistribution (Altieri 2009; Gliessman 2016; Holt-Giménez 2017).

While land scarcities and toxic soils lead a list of obstacles to upscaling urban cultivation, there are others, including the environmental impacts of decentralised production and the generation of new sources of food waste (Boyer and Ramaswami 2017; Mohareb et al. 2017). Both macro- and micro-level data support the point that instead of heralding substantial output as the goal of urban food growing, it is realistic instead to cast it as a small secondary gain. Food growing has much more to offer cities and their inhabitants (Santo, Palmer, and Kim 2016). This includes bolstering social sustainability through two primary vectors: countering food inequality and promoting public health. For food output, the essential tasks remain in the realm of agriculture outside cities. They include meeting a number of formidable challenges, such as shifting field crops away from animal feeds and biofuels (Cassidy et al. 2013)

Ecological sustainability?

Urban agriculture's bandwagon promises contributions to ecological sustainability, and there is more empirical support for this message than the one for food security (much less for food access, which is rarely addressed, if at all). Many claim that urban food growing promotes ecological sustainability by providing natural habitats and biological diversity, improving soil quality (and even building new soil), reducing soil erosion, sequestering carbon, conserving biological diversity, and mitigating the heat island effect (Bousse 2009; Palmer 2018; RCEP 2007; Wolch, Byrne, and Newell 2014). There is also a possibility that it provides habitat for bees, which may contribute to pollinating surrounding plant life (Langellotto et al. 2018).

Regrettably, there is also considerable reason to question these claims and findings. To recapitulate part of the discussion from Chapter 3, urban cultivation is good for some organisms more than others, and it might promote more biodiversity but not necessarily inclusive of native species (who may lose out). Carbon sequestration hinges on gardening practices, so it is not a given, especially compared to other kinds of green spaces. Large patches of grass or, even more effectively, trees, could promote sequestration, but this also depends on what sort of trees. Improvement in soil quality and erosion reduction and in attenuating urban heat island effects may be less controversial claims. Nevertheless, the matter hinges on where gardens are located and how extensive an area is relative to local paved areas. It also depends on how large an area is cultivated, as it might make little to no difference overall when a city's total size is considered.

There are also major questions about expanding such ecological contributions. The primary one is that the upscaling needed to convert present urban green spaces (greenbelts, forests, parks, etc.) to growing food will detract from sustainability for two reasons. Firstly, these green spaces already provide for the ecological benefits. Secondly, conversion of green land to food production will entail considerable energy and other ecological drawbacks (see Chapter 6). This makes for a case-specific and uneven profile with regard to the environmental benefits and losses of the change (Fisher and Karunanithi 2014). These profiles would require systemic analyses such as life cycle assessments (LCAs) in the choices of food-growing sites and the materials to be used. Such assessments focus on GHG emissions as a sustainability marker that includes mostly CO_2, as well as methane and NO_2, which affect radiation, leading to warmer climates (FCRN 2015, 3–20). It is not the case that urban food-growing plots can be assumed generically to contribute to ecological sustainability. For example, Kulak, Graves, and Chatterton (2013) carried out an LCA of a community farm in London and found that reduction in emissions come *only* from an appropriate choice of crops—those that can substitute for foods with high GHG footprints. At the same time, LCAs do not address societal impacts, which include provision of ecological and environmental knowledge and experience and creation of neighbourhood social networks that can serve to strengthen climate change resilience in cities (McIlvaine, Porter, and Delany-Barmann 2019).

The argument most commonly made for the ecological sustainability contributions of urban food growing is its localness. The gist of this point is that locally sourced food requires less transport energy and thereby lowers GHG emissions—the food miles argument (see Turner 2011 for a review). However, the argument has been challenged by an ample body of analyses and studies (CCAFS 2012; Meyer and Reguant-Closa 2017; Morgan 2009; Peters et al. 2008; Stancu and Smith 2006). The critiques are based on the compelling finding that techniques used to produce food far outweigh transport in its life cycle impacts on the environment. In the US, food processing accounts for 83 per cent of its GHG emissions (CFF 2017). Transport is responsible for just 11 per cent, the large majority of which is from local lorries' high emission levels. Global food transport is dominated by reliance on large ships with quite low GHG emissions per unit of food weight. The food miles argument may be relevant to individual cases, but these require LCAs of their specifics. It cannot be assumed that less transport from point of production to point of consumption leads to a better green profile. Foods have variable and complex life cycles relative to ecological sustainability.

Finally, with regard to both food sustainability and security, organic farming presents some generally unrecognised challenges. While it meets consumer desires for food that is not chemically treated (often with highly toxic compounds) in its life cycle, it has yield and cost disadvantages compared to conventional farming, with some exceptions (Muller et al. 2017; Reganold and Wachter 2016). One study has calculated that any widespread growth in conventional organics' current 1 per cent of agricultural production could cause a considerable loss of natural habitats and would retain high retail prices (Meemken and Qaim 2018). The problem can be surmounted if agroecological practices were to be widely adopted, but the economic variables within a capitalist system would continue to make organic foods less affordable to poor persons in the global North and to the great majority of people in the global South. With regard to food security, the cited researchers' scenario analysis showed that organic agriculture could feed the world in 2050 without habitat loss—*if* diets became entirely vegan. Hence, the problem is also far from technical. It is social and especially political in character, ranging from struggles over land distribution and farmers' working conditions to food access, its profitability and marketing pressures, and dietary patterns. In other words, capitalist relations are the biggest obstacle to any technically feasible ecological sustainability potentials of urban cultivation, just as they are to agriculture generally.

Technological output boosts?

A major means by which the urban agriculture bandwagon expects to upscale output is through technological fixes that outflank the lack of land, while at the same time largely ignoring their biophysical and ecological implications. The fixes have been described as "ZFarming" for zero acreage and are twofold: indoors in high-rise buildings and underground chambers, and outdoors on building rooftops and walls (Specht et al. 2014).

Indoors

These technology-intensive schemes do not meet the basic sustainability tests (Hamm 2016). Growing food in high-rise buildings (Despommier 2009, 2010, 2017; Martin 2016) or in underground chambers (Hickey 2015; Yuan 2015) would create carbon footprints much higher than conventionally grown produce (Al-Chalabi 2015). For example, it takes so much energy to heat greenhouses in France that its tomatoes have a higher carbon footprint than those imported from unheated Moroccan greenhouses—even accounting for their transport (Payen, Basset-Mens, and Perret 2015). Indoor urban food growing at even modest scales would require large amounts of energy for artificial lighting and climate regulation and would produce sizeable quantities of solid waste and wastewater. From an ecological sustainability perspective, it is a squandering of human intellect as well as natural resources to invest time into developing technologies that rule out the simple and direct use of solar energy—our single, fully renewable energy source (Hamm 2016; Schwartzman and Schwartzman 2019).

There are other unaddressed concerns in plans to grow food in tall buildings. One is taking the presence of such buildings for granted. Typical of technocratic visions, this is a way of normalising the abnormal, since most cities (even those in the global North) are not tightly packed with skyscrapers. Just constructing new ones or converting old ones will produce quite considerable ecological sustainability deficits.

Outdoors

Rooftop food growing, for one, does take advantage of sunlight and is touted as a substantial venue for urban food growing (Mandel 2013). It has the advantage of not requiring new buildings. Among the drawbacks is its very limited space, and that it does not provide ground-level, entry access to neighbours (and potential gardeners). Green roofs are also difficult to integrate into a building's waste management and recycling systems (Sanye-Mengual et al. 2014). Moreover, start-up and operating costs are high for rooftop gardens, just as they are for high-rise and underground food-growing venues. It is estimated that vertical farms would cost 1.5 times as much in per weight unit of food as traditional farms (Duggan 2018). This would likely limit their output to high-end niche crops such as lettuce for upscale restaurants.

The promotional movement for roofs centres on carbon sequestration, reduction of net energy consumption and urban heat island effects, raising moisture retention, and increasing biodiversity by providing more species habitats (Berardi, GhaffarianHosseini, and GhaffarianHosseini 2013; Li and Babcock 2014; Vijayaraghavan 2016). If food production and other vegetation and soil cover are supplemented by photovoltaic systems (where environmentally feasible), there would be even more usefulness to such rooftops (Chemisana and Lamnatou 2014). Some are so enthused as to claim that urban rooftop farming is a major path to restoring food system integrity and creating more ecologically sustainable cities (Steier 2018). Given that in the green-roof leading country, Germany, only 14 per cent of roof area has so far been converted, and notably only flat roofs (Whittinghill, Rowe, and Cregg 2013), one can be forgiven for being less sanguine about rooftop gardening replacing conventional farming techniques. Worse, at least in the case of rooftop-leading Germany, soil pH and depth vary negatively with rooftop age (that is, the older rooftops get, the thinner and more acidic soils become), and vegetation composition tends towards lower biodiversity (Thuring and Dunnett 2014).

Additional technical conundrums seem not to discourage technology-fix spirits. Green rooftops need to be lightweight so that there is a preference for volcanic materials, clays, and plastics to create technogenic substrates to grow plants (Nels 2017). This implies increased quarrying in volcanic and clay-rich areas, with relatively large outlays of energy and GHG emissions. Moreover, there is a not trifling re-direction of investments necessary to retrofit buildings (including strengthening structural support) and devising context-appropriate vegetation cover or cropping systems. The latter is especially urgent because the vast majority of studies are limited to industrialised countries mostly in temperate zones (Vijayaraghavan 2016). There are foreseeable ecological challenges. One is the possibly low cooling efficiency of vegetation in Mediterranean climates (Schweitzer and Erell 2014). Other problems are the additional water requirement in drier regions (Razzaghmanesh, Beecham, and Kazemi 2014)—especially as irrigation volume correlates with cooling effectiveness (Li, Bou-Zeid, and Oppenheimer 2014)—and the leakage of nitrogen and phosphorus as runoff cascading onto lower-lying ground. These issues can exacerbate water shortages and pollution, requiring much more attention in research and implementation. It is also unclear whether or not green rooftops enhance soil biodiversity, and the outcome so far seems negative as far as micro-arthropods are concerned (Rumble and Gange 2013).

What may very well be a clinching argument against rooftop cultivation is the issue of pollutants. The apparently very high rate of pollutant absorption (e.g. heavy metals), glorified as a major benefit, is actually a possible contamination hazard if rooftops are to be cultivated (Li and Babcock 2014; Seidl et al. 2013; Vijayaraghavan and Joshi 2014). Such findings imply questions about the runoff content from older green rooftops in countries within humid temperate areas, where precipitation tends to abound, such as in Germany (Thuring and Dunnett 2014). This is not being taken into consideration in studies such as those on the Brooklyn Grange Navy Yard Farm rooftop in New York City, where only short-term contamination from rooftop drainage and atmospherically deposited heavy metal contamination were found to be minimal (Harada et al. 2019). The same study indicated that heavy metal contaminants accumulate appreciably in soil during fallow periods (winter months). To reduce this hazard, the authors recommended a mulch or vegetation cover that should be treated essentially as hazardous waste. The longer-term potential for contamination hazard and the production of contaminated mulch or vegetable matter is noteworthy with respect to the potentially deleterious contributions of rooftop food growing. Furthermore, a generally declining pH with rooftop soil age (related to high rainfall, especially when it is acidified from industrial sources) could make many heavy metals more water-soluble and thereby contribute to more down-building contamination. Vegetable productivity on rooftops, at any rate, appears to be highly variable relative to species or variety and therefore still needs to be carefully assessed before one can conclude anything about improvements in potential sustainable food provisioning (Whittinghill, Rowe, and Cregg 2013).

On top of ecological problems, food growing on buildings faces social obstacles. In addition to excluding non-residents, buildings' green roofs generally involve conflicts over ownership and access, maintenance and financial (including insurance) responsibilities, and decision-making processes about what is to be planted and towards what ends. These issues are not being addressed adequately, if at all, in current research. In light of these problems, rooftop cultivation can at most complement other initiatives and only when clearly shown to provide net ecosocial benefits, which so far it manifestly cannot.

A second sun-available technological option is growing vegetables on buildings along vertical or altitude gradients, using water as medium or just suspending plants in air. To some extent this is another strange fascination (e.g. Beniston and Lal 2012; Dubbeling, Orsini, and Giaquinto 2017) given that the output is, as with rooftops, meagre regardless of techniques employed. Furthermore, there are not a few challenges with implementing vertical production systems, even if they were to exclude agrochemicals. First there is the need to bring enough light, water, and nutrients to the plants. This is a problem even if the engineering difficulties of soil or water loads on buildings were surmounted through aeroponic techniques (plants suspended in air, with nutrients and water sprayed). Temperature and humidity must be closely monitored and adjusted to enable crop survival and to keep pests away.

These technologies make for high-energy-demanding operations, such that it is unclear how indoor or outdoor, angular or vertical Zfarming, would make cities more biophysically and socially sustainable (Mok et al. 2014, 26). The aeroponic option, as in the case of hydroponics, fails to account for the possibility of negative effects to plant health due to the absence of symbiotic arrangements with other organisms, especially fungi and bacteria (see also Biel 2016, 97). Applying genetic modification to overcome these and other challenges is yet another example of downplaying the foundational roles of evolutionary and ecological processes. It also courts other negatives, such as the unintended proliferation of modified genetic material. More importantly still, politically, is the issue of who controls gene-modification technologies and their distribution, and the affordability or general accessibility of genetic modification to cultivators.

In summary, food growing in or on buildings lacks community engagement beyond the people inhabiting the building—that is, unless the other inhabitants nearby will be allowed to enter a building where food is produced. This can be done, as in examples of community rooftop gardens in Toronto and Lyon (Nasr, Komisar, and de Zeeuw 2017, 12–14). Under typical capitalist ownership conditions, however, this is unworkable because the community must still ultimately answer to a building owner. The scheme will not, in any case, address land speculation, houselessness, and private ownership. What instead prevails are disquisitions on the merits of municipal intervention and support for urban greening projects, or, more brazenly, the best way to give "the private sector" a hand in carrying them out (e.g. Mees et al. 2013). Proponents of ZFarming technology seem uninterested in the social ramifications of their projects and even less in addressing existing political inequalities.

What do gardeners have to say?

An unexpected finding from research is that gardeners do not seem to be on the bandwagon as those who tout the output and sustainability potentials of urban agriculture. A study in the UK indicates that urban gardeners have only a passing interest in providing food or in contributing to environmental sustainability (Holland 2004). Food output ranked sixth in a list of nine goals of the sites, and it was noted by less than one-half of respondents. Providing for ecological sustainability ranked fifth, just ahead of providing food. Findings from Mississauga (near Toronto) point to a similar disconnect between urban sustainability policy and gardener motivations, which are principally about personal benefits unrelated to wider, environmental outcomes (Conway 2016). For several community gardens in Brooklyn and Queens, Aptekar and Myers (2020) report that amenities (provision of green space) and food justice feature among gardeners. Even among home (i.e. single-household) gardeners institutionally certified for biodiversity conservation, as observed in Winnipeg, the main benefits listed were much more about individual lifestyle improvements than ecological ones (Raymond et al. 2019).

A recent review of community garden literature coupled with interviews of participants in six sites in Melbourne concluded that gardeners are not dominated by a desire to produce food; instead, "motivations for participation are diverse and span a range of ancestral, social, environmental, and political domains" (Kingsley,

Foenander, and Bailey 2019, 745). A survey of community gardeners in the US produced a comparable result. Health and social benefits were the most frequently cited and highly regarded goals of their activities (Parece and Campbell 2017). A fourth study, a qualitative content analysis of interviews, found the dominant motivation of gardeners in Berlin, Hamburg, and Munich was personal interest, including recreation and healthier eating. Environmental motivations lagged far behind (Zoll et al. 2018).

These findings are consistent with those from the aforementioned case studies in London, New York City, and San Francisco. For example, most of the gardeners interviewed reported that they grew food because they liked gardening. One reported that she just grew "some nibbles" but enjoyed using her "green thumbs." Furthermore, most community gardeners reported that they liked the cooperative aspect of their garden and enjoyed its green landscape features and its sense of solitude. A much smaller number mentioned just one environmental issue related to their garden—the re-use value of its triad of large composting structures. The broad appeal of Manhattan's WSCG is based in its floriculture, which occupies space equal to that of its food growing. In fact, growing flowers in community gardens for the purpose of providing aesthetically rewarding landscapes is a common activity. However, they represent the threat of yet another reduction in land availability for producing food. Floriculture is becoming a more common product in urban agriculture's expanding roll of entrepreneurial gardeners. For example, one such gardener in Detroit was quoted as saying that "per square foot, flowers are one of the more profitable things you can do with dirt" (Cowley 2015, B8).

The pattern of gardener motivation is somewhat different in a Chongqing case study (Rock et al. 2017; see Chapter 1). Though some interest was evident regarding recreational opportunities, gardeners overwhelmingly cited the fulfilment of subsistence needs as a primary motivation (to make ends meet) in terms of saving on grocery expenses and improving nutrition. Only three of the 37 interviewees marketed their surpluses. About a third (especially the elderly) thought gardening to be important because it enables them to be engaged in something constructive as well as to retain their cultural roots (for those coming from the countryside). These kinds of motivations, self-provisioning and health concerns, coupled with cultural affirmation, resemble those found elsewhere in the global South (Orsini et al. 2013) and possibly among those in the poorest sections of global North cities, where deprivations resulting from capitalist policies make it more likely that self-provisioning will occupy a prominent role.

What gardeners report in these studies as their principal interests in growing food—nutrition (e.g. satisfying food needs), health, education, community development, and leisure—are social, not ecological in character. This supports our contention that urban food production is more accurately described as cultivation, rather than agriculture, and that most of the conventional reasoning given to promote urban agriculture is at odds with gardeners' ideas and practices. From the beginnings of its contemporary resurgence, studies have identified the social dimensions of community gardens. For example, a 1997 field research in San

Francisco found that its gardens were "closely associated with environmental justice and equity" concerns (Ferris, Norman, and Sempik 2001, 559). In other words, the technical prerogatives of the urban agriculture bandwagon advocates clash not only with actually existing food-growing and ecological processes but also with urban cultivators' motivations and lived experiences. To draw from Henri Lefebvre (2000), in the social production of urban food, there is an evident chasm between lived and (institutionally) conceived spaces. That is to say, gardeners' understanding of what they do and want differs from the dominant viewpoints of policy makers, scholars, and researchers. For those of us interested in building ecosocialist alternatives to capitalist relations, this gap is a promising starting point. For example, it indicates the powers that be are not interested in what gardeners most value, and this underscores the importance of supporting what community gardeners are doing. Gardening practice can promote the development of a biophysically informed understanding and outlook. This is another facet of what we envision by urban cultivation, which is already made possible thanks to existing initiatives and practices by many urban cultivators.

Urban food growing as cultivation

It is, in part, because of its occasional role in post-industrial redevelopment that urban food growing in cities of the global North has experienced a status upgrade. It has become a *demi* movement that is in transition from being gardening to re-branding as agriculture. This shift has been described in these words,

> As we find ourselves once again in the throes of a crisis of capitalism, the popularity of UA [urban agriculture] in the global North has surged and the discourse surrounding it has shifted from one of recreation and leisure to one of urban sustainability and economic resilience. Even the terms used to describe it have shifted in the global North; "urban agriculture" is replacing "community gardening" in everyday parlance' placing it (despite its much smaller scale) in the same category as UA in the global South.
>
> *(McClintock 2010, 191)*

Urban agriculture, in another analyst's words, "has become integrated into an ideological movement of environmentally and socially sustainable choices, community networks, reconnections with nature, and social change in North America" (Mok et al. 2014, 25).

Cultivation is a term that better captures its sensibility and, significantly, its potential contributions to social sustainability. Agriculture has a long lineage connecting it to fields and pastures, as pointed out in Chapter 1. To cultivate also means to educate and to develop. It signifies and highlights the fact that urban food growing is a social practice, in which gardeners participate in community life and create new forms of urban space. In this sense, agriculture is a noun while cultivate is a verb. At the bottom, urban agriculture denotes an unattainable aspiration and cultivation speaks to a real possibility. Cities have ample space for cultivation but not for agriculture. The general use of the term urban agriculture implies an exaggerated claim of output potential. However, production is extremely uneven across different cities, mainly focuses on vegetables, and has negligible impact on total food output. Anyway, as food justice activists have argued, greater food security is not gained simply by producing more food—it is about access to food, as discussed in Chapter 1 (Cadieux and Slocum 2015; Carlisle 2013). Equitable access to food is sometimes referred to as "food sovereignty"—that everyone, even the poorest of us, have a measure of agency over what we eat (Mark 2020). Urban agriculture is often left unspecified or defined so broadly (Mougeot 2006, 82) as to make it impossible to distinguish, for example, subsistence from hobby gardening. Sovereignty over a hobby garden or a small patch of city land will scarcely help overcome the systemic and massive food access injustice intrinsic to capitalist relations.

Adding to such ambiguity and confusion is an unacknowledged problem with the spatial units used to assess urban agriculture. Cities are not clearly demarcated phenomena over the Earth's surface, so one cannot easily distinguish intra- from peri-urban food production. Urban areas grade into rural hinterlands. It is a rather splotchy curvilinear relationship, not a concentrated linear one. Moreover, there is a tendency to omit inter-linkages among cities as well as those between urban and rural spaces (Angelo and Wachsmuth 2015), thus treating urban agriculture as somehow isolated from regional and global food growing, both in its social and biophysical characters (Engel-Di Mauro and Cattaneo 2014; McClintock 2010).

The recent and largely prescriptive City Region Food Systems approach could be helpful in this (Blay-Palmer et al. 2018). The stated objectives revolve about finding potentials for systematically strengthening, if not creating, mutually beneficial linkages among rural and urban dwellers such that food access is secured to all in ecologically sustainable ways. These would be fine ideas if policies were not beholden to capitalist prerogatives. So far, however, the proponents evade any critiques of capitalism, find no contradiction in marrying profitability with food provisioning, and offer no analysis of and strategies to undo relations of domination. They opt instead for a blissful "multiple stakeholders" strategy, as if everyone could partake of decision-making processes on an equal footing in a grossly and systematically unequal society. Emblematic of this viewpoint is reference to highly problematic urban planning examples from cities like Colombo, Sri Lanka, and Rosario, Argentina (see Chapter 5). It is therefore unsurprising to find proponents of this approach invoking UN and EU policies while ignoring constructive experiences among socialist governments, which provide a much more effective grounding. If one is really interested in raising standards of living and livelihoods across urban and rural areas, while promoting ecological sustainability, then it should be useful to have a look at successfully working precedents, as in Cuba.

At the end of the day, relative to what goes on in cities proper, while urban food growing is important, it is not for its output or its ecological sustainability, rather, because of the multi-faceted benefits it provides for communities (WinklerPrins 2017, 3). In some measure, urban food growing may have become designated as agriculture because it shared the public stage with a simultaneous surge in local foodism, which challenged rural agriculture's faults with an urban version's promises.

Foodism and localism

Urban agriculture's growth spurt in the global North has led some planners and designers to re-imagine the city itself as a farm—a landscape that is continuously food-productive (Viljoen 2005). The urban-centred food production landscape generally follows a pattern of (1) inner urban or core community gardens; (2) suburban domestic gardens; and (3) peri-urban farmland tracts (*fresh* 2012, 9). It is an indication that localisation has come a long way since its emergence in the new social movements of the 1960s with their message of distrust for large-scale, centralised organisations (Allen 2009). Neo-localism (Schnell and Reese 2013) or re-localisation (Hein, Ilbery, and Knefsey 2006) has gained ascendancy in large measure behind condemnation of the standardised products and placeless landscapes of the agri-business economy (Morgan, Marsden, and Murdoch 2006).

Localism is defined as the building and maintenance of personal attachment to a place or locale. It is important to note that localism is not limited to the global North. For example, it is a documented phenomenon in Brazil (Klink et al. 2011) and in Thailand (Southard 2014; Parnwell 2007). It also features prominently in global North nations other than the US. In the EU it has "led in many countries to the emaciation of regional policy to be replaced by a neo-localism" (Boylan 1997, 631). However, localism is a social and political strength without recognition of its interconnectedness with other levels of life and struggle. Rather than "think globally, act locally," the better mantra is "think and act, globally and locally."

Popular trends in the global North have rebelled against distant industrial food production in favour of alternative local and artisanal food growing. They have stimulated the surge in urban agriculture. An early benchmark in this rebellion was a new focus on ingredients rather than food preparation techniques. This led to the fresh, farm-to-table model for food acquisition. Californian cuisine, as prepared by Alice Waters' (2011) in Berkeley since 1971, is an eminent exemplar. The Slow Food trend, which emerged in Italy's Southern Piedmont region in 1986 and uses traditional culinary practices, has become another headliner of local food production and consumption (Andrews 2008). It directly counters the prevailing assembly-line standardisation of fast food (Schlosser 2002). In sum, an apt description of the local foodist movement is that it encompasses everything about eating, including what we consume, where it originates, how we acquire it, and who prepares it (Lawrance and de la Pena 2012, 2).

The social and cultural trends of localism and foodism are so coupled as to be referred to in the same breath, as local-foodism. Perhaps in no other new phenomenon is this close relationship manifested than in micro-craft breweries (Holtkamp et al. 2016; Pole and Martin 2017; Schnell and Reese 2014). It is a rapidly growing part of the cultural turn towards local and artisanal production. Like artisanal food, micro-craft beer has become a vessel through which people can engage with local places (Graefe, Mowen, and Graefe 2017; Kline, Slocum, and Cavaliere 2017, 13). Beer and food have become means by which individuals can construct a self-identity within a locale and at the same time can signify a social status within a community. A common way to promote and to take marketing advantage of localism is by branding craft beers with names and events from local history and with prominent local landscape features (O'Neill, Houtman, and Aupers 2014). Significantly, and not surprisingly, because of their higher costs, local, fresh beer and food sold in brewpubs have a largely white and middle-to-upper class base in the US (Murray and O'Neill 2012).

Local-foodism is commonly promoted under the appellation of locavorism. The term was inaugurated in 2005 to describe the practice of eating a diet composed of food grown within a 100-mile radius, and it was the *New Oxford American Dictionary*'s word of the year in 2007 (Ladner 2011, 11–12). Locavorism's popularity has benefitted entrepreneurial start-up enterprises, often digitalised, in the lucrative market generated by upper- and middle-income households in better-off urban areas able and willing to pay a premium price for fresh and local food (Bosco and Joassart-Marcelli 2017). Thus, an analysis of North American urban agriculture concluded that it is more often practised in high-income neighbourhoods. Low-income ones are more likely to lack access to fresh produce, green space, and other benefits (Gray, Diekmann, and Algert 2017).

California epitomises the production duality of present-day rural agriculture in the global North. It is the leading food-producing state in the nation that is the world's largest exporter of food (FAO 2013). Thus, it has been at ground zero for both the industrialisation of agriculture and the local-foodism trend. The state grows a massive level of relatively inexpensive foods on large farms that is exported globally as well as marketed nationally. A rather extreme example of Californian agriculture's scale is almonds. It is the only US state that grows them for commercial purchase, and it produces over 80 per cent of global output. It also features a contrasting, more intimate, and much pricier output from small-scale producers who sell to high-visibility farm-to-table outlets, especially in its large urban agglomerations of Los Angeles and San Francisco (Starrs and Goin 2010, 51).

From an analytical perspective, local-foodist practice is based in reasoning that contains as much wishful thinking as realistic assessment. Overreach about its features is prominent in its discourse. An illustrative and somewhat grandiose example is a list published by the University of Vermont that includes ten reasons to buy locally grown food: it tastes better, is better for you, preserves genetic diversity, is safe, supports local families, builds community, preserves open space, keeps taxes down, benefits the environment and wildlife, and is an investment in the future (Grubinger 2010). It is possible that the drumbeat reporting of such grand accomplishments puts off some people as well as changes the habits of others. In the extreme range that exists in personal habits, local-foodism can become an individual's totem, as is the case with many commodities. The local food fetish has been labelled by some as an eating disorder—*orthorexia*. It features an obsessive behaviour pattern of seemingly endless pursuit of the healthiest locavore diet (McWilliams 2009; Scelzi and La Fortuna 2015).

The optimism of local-foodism is encountering a fresh scepticism. The reality is that local does not equate with sustainable-production-by-nearness. Even food that is designated as local is often consumed hundreds of kilometres from its production site, and it is not always sustainably grown and so certified. Perhaps the most widely disseminated critique has been the "local trap" analysis by Born and Purcell (2006). It argued that urban food growing is based in an uncritical assumption that local is inherently, *de novo*, preferable to other production venues. Its optimistic assumptions sweep across a whole compass of achievements—more urbanites fed and ecological sustainability gained, as well as the popular themes of more nutritious, fresher, and better-quality food (Johnson, Aussenberg, and Cowan 2013).

The accumulating critical studies of local-foodism focus on lack of empirical evidence for the optimism about its agricultural and sustainability potentials, and on its shortcomings in dealing with food-connected injustices that are present in the life cycle of food. Moreover, there is much evidence showing such an approach in cities like Dar es Salaam is not only inappropriate, its application would worsen already high rates of malnutrition (Wegerif and Wiskerke 2017). What remains underdeveloped in these analyses is a progressive, place-sensitive political ecology to re-frame urban agriculture going forwards. Without such re-framing, local-foodism will continue to play into neoliberalism's private-market assault on public regulation and the commons (DuPuis, Marsden, and Murdoch 2006; Hess 2009). It can begin by making whole the truncated social leg of the wobbly sustainability stool. While the movement is at least in part well-meaning, it starts from bourgeois premises, such as the notion of changing society just by changing purchasing patterns and by maximising the parameters of individual choice.

Social sustainability

Global diffusion of the term sustainability, as it relates to environment, began with a UN (1987) plenary resolution. It has grown to become a cultural meme serving as the sign of a new human behavioural system of thought and practice, spurring a wealth of popular and technical literature. Its status has grown even more as a result of the emergence of threats that climate change poses. From the beginning, social development has been the most amorphous and least considered of the three recognised vectors of sustainability practice and research—environment, economy, and society. The economy has been the environment's adjutant in the workings of sustainable development policies and programmes. The UN resolution put an emphasis on economic growth ("development") as the basis for sustainability. This focus on growth has been subjected to a growing criticism in ecological economics about its contradictions and limits (Hirsch 2005; Jackson 2009). Furthermore, radical reformist approaches to neoliberal capitalism have produced neo-Marxist ecosocialist critiques of growth (Leahy 2018).

Social sustainability is described aptly as a "concept in chaos," a circumstance created in part by official inattention:

> Though the concept of sustainable development originally included a clear social mandate, for two decades this human dimension has been neglected amidst abbreviated references to sustainability that have focused on environmental issues, or been subsumed within a discourse that conflated "development" and "economic growth."
>
> *(Vallance, Perkins, and Dixon 2011, 342)*

At present, society is the neglected sibling in the family triad of sustainability. The urban agriculture bandwagon is no exception due to its focus on food output. The lack of consideration for society and its injustices, inequities and inequalities, is consistent with a hegemonic neoliberalism dominated by the demands of the private sector for economic growth and business profits (Giddings, Hopwood, and O'Brien 2002).

The urban food-growing sites examined here in the London, New York City, and San Francisco study (Martin, Clift, and Christie 2016) support a new perspective on the current widespread enthusiasm for urban agriculture. While structural limits will prevent urban food growing from reaching the status of agriculture and while its contributions to ecological sustainability are uncertain, there is a strong case to be made for its contributions to social sustainability. The three sites indicate that urban food growing can produce little more than "nibbles" of food but it can contribute "oodles" of social sustainability services. It appears that gardeners are on the same page with regard to the purposes of their plots, as remarked earlier. In the studies cited earlier, as well as the cases examined here, the leading goals submitted by gardeners were fundamentally social in character—education, community engagement, health, and recreation.

The range and variety of urban food-growing modalities illustrate the primacy of the social rather than the ecological or agricultural in their activities. There are community gardens, allotments, small and large farms. There are school gardens, domestic gardens, prison gardens, and plots to train aspiring farmers. There are plots with work-study programmes and community services for disabled persons and others, including school children (Desmond, Grieshop, and Subramaniam 2004; Subramani and Selvan 2014). The list continues to grow. An inventory of urban food growing's multiple formats in North America organised them into six categories: home gardens, community gardens, non-profit urban farms, for-profit urban farms, institutional gardens, and interstitial food spaces (Gray, Diekmann, and Algert 2017, 24). There are also cross-over plots—for example, a food-growing community hub that operates as a social service organisation in inner city Vancouver (Ableman 2016). It is instructive to note that the various community or city farms involved in urban food growing are not the traditional private undertakings. A study of an urban farm in Baltimore found that it welcomed their neighbours to participate in publicly sponsored programmes and provided them with employment opportunities as well (Poulsen, Neff, and Winch 2017). While differing in specifics such as size and locale from community gardens, the urban farms are actively engaged in neighbourhood development.

In its support of community organisation, urban food growing makes meaningful contributions to two major components of social sustainability: environmental justice and public health. Both are needed now more than ever—environmental justice because of dramatically widening inequalities (Coote 2014), access to a healthy diet being one, and public health because of the contemporary obesity epidemic in the global North (Freund and Martin 2008). Obesity has been designated as an underlying condition that promotes infection and deterioration by COVID-19 (CDC 2020). Even a very small food-growing space can contribute to environmental justice. An apt example is a half-acre (0.2 ha) community garden in one of the poorest neighbourhoods in New York City. It employs a dozen teenage boys with criminal records to grow serrano peppers, working under court orders as an alternative to incarceration. Their small stipends come from selling the garden's *Bronx Greenmarket Hot Sauce* (Winnie 2015).

However, given the possibility of high contamination levels in New York City, it seems that gardeners may be putting themselves at greater contaminant exposure risk or adding to existing forms of exposure, like water and air pollution (see Chapter 3). Politically, this calls for accountability that cannot but be socialised and hence cities must become (health) commons—because of the absurdity of trying to socialise accountability while maintaining private profits. As to other issues of public health, there is abundant evidence of the ways in which all urban green spaces contribute to physical, psychological, and social health (Abramovic, Turner, and Hope 2019; Brown and Jameton 2000; Cattell et al. 2008; Comstock et al. 2010; Ferris, Norman, and Sempik 2001; Golden 2013; Litt et al. 2011; Louv 2008; Poulsen et al. 2014; Pugh 2013; Relf 1992; SDC 2008; Silva 2016; Weinstein et al. 2015; Wolch, Byrne, and Newell 2014). With regard to humanity's affinities with nature there is a hypothesis, *biophilia* (Wilson 1984), which posits that our species has an innate tendency to experience connection with other life forms, including flora. Urban food growing has been offered as an example (Beatley 2011), but to bring such a tendency to prominence takes the overcoming of alienation from life brought about by capitalist relations.

Perhaps the most widespread social sustainability contributions of urban food growing are in community development and education (dietary and environmental). This may occasionally include the facilitation of cross-cultural exchange to diffuse more sensibility about the struggles against settler colonialism in North America by engaging neighbourhoods with substantial First Nations, for example (Datta 2019). Then again, in places like Copenhagen, the process of strengthening social capital seems to deepen pre-existing bonds among Whites, rather than fostering inter-cultural understandings (Christensen, Malberg, and Allenberg 2019). An analysis in the UK found that a sense of community participation and empowerment features in community gardening (Holland 2004). Food growing can enhance the creation of neighbourhood social capital built through networks of human relationships. Recent studies of UK food schemes have reinforced this point and identified a range of social networks, for example, in developing informal research and demonstration projects (Durrant 2016; White and Sterling 2013). A study of community gardens in Nottingham found that they helped to build community cohesion and vitality and were both a consequence of and a source of social capital (Firth, Maye, and Pearson 2011). Social capital has yet another significant distinction—it is the essential similarity shared between urban food growing in the global North and South (WinklerPrins 2017, 3).

Such findings about social capital underscore a key point in our critical analysis of urban food growing—it is primarily about the *cultivation* of social skills and capabilities and, potentially, biophysical science and like technical skills, that is, a largely educational endeavour. Thus, for example and to reiterate, a literature review of urban gardening studies concluded that it supplies only a small amount of food and that its contributions to environmental sustainability are at best debatable (Santo, Palmer, and Kim 2016). Social sustainability services promote public health in all its facets and reduce inequalities in all their manifestations. They are quite commonly understood but also quite difficult to identify and assess, unlike the metrics that are available for food output and ecological sustainability. Efforts are underway in research such as the Social-LCA (Kuhnen and Hahn 2017) and in policy such as the UN's Agenda 21 to develop suitable sets of empirical indicators for valuing social sustainability benefits (Beilin and Hunter 2011). Such social indicators would be useful information as an accessory to the social changes that are necessary.

It is in education that the above-discussed case studies in London, New York City, and San Francisco make their most impressive contributions to social sustainability. This is significant because the inter-generational, first principle of sustainability relies on providing for environmental education for new-age cohorts. The community garden in New York City reserves six plots for school children to participate in an ecology learning module during which they grow and eat vegetables. It is noteworthy that most of the adult participants in the garden have had previous gardening experience in their childhoods. As a follow-up to their experiences, children and their teachers have constructed several raised beds in their nearby schoolyard. In New York City, the number of registered school-based gardens has multiplied sixfold (Foderaro 2012). The small suburban farm in London operates a funded school programme in which pupils and staff, after school, grow, cook, and eat vegetables. In addition, students and their caretakers from a local school participate in a sponsored Disabled Farming Assistance programme.

The larger exurban farm near San Francisco, like the garden and the small farm, operates environmental education programmes for school pupils. In an increasingly important aspect of education, it also provides a rare learning opportunity for city-raised young people. Increasing urbanisation has progressively reduced the number of persons with farming knowledge. In 2012, the average age of US farm operators was 58.3 years, up 1.2 years from 2007. The declining number of beginning farmers continued a 30-year downward trend (USDA 2014). This loss of farming expertise threatens overall food production and climate change resilience. The San Francisco metro area farm addresses both issues by inhibiting the conversion of farmland to settlement and by training a new urban generation of aspiring farmers. The small farm in London addresses the same issues by providing a sustainable farming apprenticeship programme.

In addition to structured learning, informal education is part of urban food growing—for example, in the serendipitous sharing of knowledge, experience, and work among its practitioners. This has also been found to occur in Chongqing between experienced and new gardeners (Rock et al. 2017); skilling in crop production is being diffused by means of informal sharing of expertise and knowledge (see also Chapter 7). Gardeners hear from each other about some of the complexities of production and its relationship to sustainability. This communal learning is an example of the synergies that exist between ecological and social sustainability (Martin 2013) and supports the argument here that small urban food plots mainly have social value but with positive ecological consequences because of their potential for raising awareness about ecosystems. The communality of learning results in an accumulation of social capital—of which the most needed application is to community and organisational leadership. This tendency for social capital building should not be confused with any necessarily egalitarian or communalistic impulse or prospect. As stated earlier, existing critiques of urban agriculture have shown that it can be a highly racialised, gendered, and heteronormative reality that often reinforces existing relations of power.

Nevertheless, as growing food is a thoroughly embodied experience (Freund 2008; Turner 2011), urban cultivation may have more learning impacts in a person's life than do other environmental education activities. Social learning takes place in urban food growing through a sharing of ecological observation and monitoring by gardeners (Irvine, Johnson, and Peters 1999). The learning often goes beyond one's green thumbs. Experiential learning can stimulate change in lifestyles. Food growers may gravitate to healthier diets (with more vegetables), and they may also take up sustainable practices such as composting. For example, a goodly number of the New York City's members discussed earlier regularly carry bits of food waste from their apartments to its compost bins, whether or not they have plots to tend. In so doing they have encouraged non-members in the neighbourhood to do the same.

Food and uneven development

To return to the local-foodism problem, in addition to problematic output and uncertain green profiles, the practice *for its own sake* is a political dead end or is a political means to reinforce capitalist relations by reproducing them, as it displaces attention from more politically charged urban food issues. De Lind (2011, 273) has argued that it helps to shift public focus away from concerns of equity and citizenship. Hess (2009, 95) makes the following succinct observation: "When evaluated from the perspective of the contribution to building a more just and sustainable society, localism is in many ways a bundle of contradictions." Research demonstrates that local food is equally likely to be just or unjust because "localities embody material and power asymmetries" (Allen 2009, 303). Local food certainly *can* be sustainably grown in the ecological sense but that is a matter for research to demonstrate and for regulation to ensure (preferably regulation from below). The fact is that the devil lies in the details: food justice as well as food quality and sustainability are not determined by scale or by location but instead by how the growing actually functions at whatever magnitude and in whatever location.

Capitalism 3.0 is a global and multinational neoliberalism which features deregulation of the private sector and outsourcing of government activity to private for-profit contractors. In the last decade, its contradictions and austerities have led to a heightened crisis situation. While there are many manifestations of this crisis around the world, its headline events have been the financial meltdown of 2008 and the twinned reactionary political surprises of 2016—Brexit in the UK and Trump in the US. A further calamity in 2020, COVID-19, is discussed in Chapter 8. The virus' aetiology and impacts are related to the global food systems of agribusinesses (Gunia 2020; Wallace 2016). Food's social justice quality, as in an access to a healthy diet, is facing increasing challenges in the era of expanded austerity. Austerity (e.g. structural adjustment) has been the mainstay for most of the world economy since 2008. However, just because it is forced on relatively wealthier countries does not mean it is universal; the changes witnessed in China being one example. There, the term austerity might not be resonant or relevant to large numbers of people.

The simultaneous growth of urban food deserts and impoverished rural communities in the global North illustrates the ongoing inequities generated by neoliberal government policies. At present, urban food growing presents a quite ambiguous political profile marked by positive and negative relationships to neoliberalism. Local-foodism supports some progressive goals with regard to environments and communities that challenge neoliberal orthodoxy (Davolio and Sassatelli 2009; Johnston and Baumann 2009; McClintock 2014). These include challenges to the products of corporate agri-business. However, neoliberal policy also gets implicit support in local-foodism's studied avoidance of political issues such as food workers' labour rights and racial-ethnic discrimination in fair access to food. With respect to the latter, a study of US farmers' markets found that a community's percentage of Black and Hispanic residents was negatively associated with its per capita number of farmers' markets (Singleton, Sen, and Affuso 2015). An analysis of the mainstream of the US urban alternative food movement concluded that it benefitted privileged white communities while its efforts to reach out to disadvantaged neighbourhoods produced as much public relations information for its sponsors as it did a lasting food justice for the deprived (Broad 2016, 197).

A major food injustice that community gardens reflect relates to the social class divisions among different gardens. The ongoing and persistent pressures of land developers and city housing authorities result in unequal land seizure outcomes. The gardens worked by the poorest citizens are the most vulnerable (Hess 2009, 143). In affluent or gentrifying neighbourhoods, community gardens are protected because of their enhancement of property values (see Chapter 2). The apparent contradictions between a dual-pronged regressive and progressive urban food growing have been highlighted in an analysis by McClintock (2014). On the progressive side, it challenges the received formats of neoliberal food production and consumption (Lyson 2004). On the negative side, it values private self-provisioning, which provides an avenue for governments to further reduce public food support programmes (WinklerPrins 2017). Trying to straddle the progressive–regressive divide has proven to be a fraught undertaking. A study of urban food growing in Melbourne found that the use of both insider and outsider tactics by local-foodism advocates resulted in various frustrations (Lyons et al. 2013, 162). A growing consensus of critics argues that local-foodism commonly supports the principal goals of neoliberalism: deregulation and more dependence on private markets (Weissman 2014). One suggested change in its ambivalent politics is to forge alliances with groups that address issues of equality and justice in order to transcend its presently unreflective localism (DuPuis and Goodman 2005).

Neoliberalism is not omnipresent. For example, there is evidence that municipalities can bypass national governments and acquire support from other institutional sources. For example, a study in Naples examined the workings of an FAO-sponsored programme in which poor and unemployed persons could improve their socioeconomic positions through urban food growing (Rusciano, Civero, and Scarpato 2017). On a macro level, the political ecology of rural production and urban consumption of food illustrates the Marxist concept of uneven development, particularly as it has been elaborated and applied to the contemporary world by Lefebvre (2000) and Smith (2008). Essentially, uneven development refers to the socioeconomic disparities that are propagated in the interest of capital accumulation, resulting in large inequalities within and between countries, as well as between and within cities. Within countries in the global North, uneven development is reflected in new levels of inequality produced by neoliberal austerity regimes that roll back or prevent adoption of new public policies to support the growing numbers of socially disadvantaged and excluded persons. This inequality is reflected in the micro-level workings of urban food growing. One can point to the already cited study of a food-growing project in New York City (Cohen and Reylolds 2015), where the researchers found significant socioeconomic disparities among gardeners in their access to both public and private resources.

In part, the local-foodism trend and urban agriculture represent progressive efforts to reassert individual identity and subgroup heterogeneity in our increasingly globalised, standardised, and outsourced political economies. However, particularistic identity politics serve to replace universal class politics, and urban bourgeois food consumers remain divorced from rural working-class field hands, many of whom are poor immigrants from poor nations. About nine-tenths of California's large farmworker population is foreign-born, primarily *campesinos* from México. Despite working in the nation's agricultural salad bowl, these workers suffer much higher rates of obesity than the national population, largely due to their overconsumption of cheap sugary drinks and high-calorie fast foods (Fuller 2016).

Meanwhile, the fruits of their labour go to pricier outlets in distant better-off urban neighbourhoods.

Neoliberal austerity fetishises price-restricted private markets over democratic commons or public goods, adding to both urban and rural socioeconomic inequalities. The accumulation and intensification of inequalities has energised reactionary populist, racist, and nationalist politics, dramatically signalled globally through the diffusion and increasing election successes of extreme right-wing parties. It is indicated in the policies of India's Modi and in other political regimes around the world (e.g. Hungary, Poland, and Turkey). In the US, right-wing resurgence has led to the Trumpian wall with México; the UK's withdrawal from the EU is another version of walling-off immigrants. Neoliberal austerity is now facing a worldwide confrontation with the consequences of its policies. It is demonstrating its contradictory character, and governments are now striving to re-direct the political fallout of their repressive and destructive policies away from themselves and the capitalists they represent and towards scapegoats, be they Muslims, migrants, or people of colour generally. However, backward-looking right-wing populism is losing its ascendancy in some countries, including the US. The BLM multinational socio-political movement, which ballooned in 2020, is shaking the historic foundations of reactionary racism (see Chapter 8).

Local-foodist discourse short-changes political struggles for self-determination, for an end to food injustices, and for progressive responses to escalating political conflicts (such as those involving immigrants and refugees) emanating from the global North–South fracture (Allen 2009; Davolio and Sassatelli 2009), and from within the nations of both. Instead, the focus is on the quality of food consumption itself and on new vehicles (and digital applications) for bringing upscale niche produce to city outlets and homes. Corporations and entrepreneurs play a leading role in privatising the economic benefits emanating from the expansion of local-foodism (Litzky, Andersson, and Smith 2017). In such ways, urban agriculture, as practised, continues on the risky path of being an instrument for supporting neoliberalism (McClintock 2015) and capitalist relations more broadly.

In an evaluation of the strengths of global food movements, Holt-Gimenez and Shattuck (2011) suggest that there is an opening for alliances between progressive and radical participants that can challenge the corporate food establishment and its neoliberal reformist supporters at the same time. Re-framing and re-directing urban agriculture to an urban cultivation designation provides an opportunity for developing a forward-looking political ecology of food. Community gardens provide an opportunity for re-visioning the historic commons in today's urban life. In an analysis of the gardens in New York City, Eizenberg (2012) makes the case that their exercise of a defiant right to public space represents an expression of Lefebvre's (2000) "right to the city." Urban cultivation, then, offers a model of the kind of social relations and local politics that can recreate a contemporary public commons that challenges the neoliberal state's hegemony. However, that challenge currently lacks a direct and organised critique of neoliberalism or of capitalism in general, a critique that would promote a movement that carries a progressive message of social justice and respect for life-supporting ecological relations. The next chapter constructs the foundations for a commensurate ecosocial approach—a cultivation that is mindful of both social and biophysical aspects of the city—that is necessary for the development of a politics to counter and challenge capitalist practices.

References

Ableman, M. 2016. *Street Farm: Growing Food, Jobs, and Hope on the Urban Frontier.* White River Junction, VT: Chelsea Green.

Abramovic, J., B. Turner, and C. Hope. 2019. Entangled Recovery: Refugee Encounters in Community Gardens. *Local Environment* 24(8): 696–711.

Al-Chalabi, M. 2015. Vertical Farming: Skyscraper Sustainability? *Sustainable Cities and Society* 18: 74–7.

Allen, P. 2009. Realizing Justice in Local Food Systems. *Cambridge Journal of Regions, Economy and Society* 3: 295–308.

Altieri, M. 2009. Agroecology, Small Farms, and Food Sovereignty. *Monthly Review* 61: 102–13.

Andrews, G. 2008. *The Slow Food Story: Politics and Pleasure.* London: Pluto Press.

Angelo, H., and D. Wachsmuth. 2015. Urbanizing Urban Political Ecology: A Critique of Methodological Cityism. *International Journal of Urban and Regional Research* 39: 16–27.

Aptekar, S., and J.S. Myers. 2020. The Tale of Two Community Gardens: Green Aesthetics Versus Food Justice in the Big Apple. *Agriculture and Human Values* 37: 779–92.

Badami, M.G., and N. Ramankutty. 2015. Urban Agriculture and Food Security: A Critique Based on an Assessment of Urban Land Constraints. *Global Food Security* 4: 8–15.

Beatley, T. 2011. *Biophilic Cities: Integrating Nature into Urban Design and Planning.* Washington, DC: Island Press.

Beilin, R., and A. Hunter. 2011. Co-Constructing the Sustainable City: How Indicators Help Us 'Grow' More than Just Food in Community Gardens. *Local Environment* 16: 523–38.

Beniston, J., and R. Lal. 2012. Improving Soil Quality for Urban Agriculture in the North Central U.S. In, *Carbon Sequestration in Urban Ecosystems*, edited by R. Lal, and B. Augustin, 279–313. Dordrecht: Springer.

Berardi, U., A.H. GhaffarianHosseini, and A. GhaffarianHosseini. 2013. State-of-the-Art Analysis of the Environmental Benefits of Green Roofs. *Applied Energy* 115: 411–28.

Biel, R. 2016. *Sustainable Food Systems: The Role of the City.* London: UCL Press.

Blay-Palmer, A., G. Santini, M. Dubbeling, H. Renting, M. Taguchi, and T. Giordano. 2018. Validating the City Region Food System Approach: Enacting Inclusive, Transformational City Region Food Systems. *Sustainability* 10: 1680.

Born, B., and M. Purcell. 2006. Avoiding the Local Trap: Scale and Food Systems in Planning Research. *Journal of Planning Education and Research* 26: 195–207.

Bosco, F.J., and P. Joassart-Marcelli. 2017. Gardens in the City: Community, Politics and Place in San Diego, California. In, *Global Urban Agriculture*, edited by A.M.G.A. WinklerPrins, 50–65. Boston, MA: CABInternational.

Bousse, Y.S. 2009. *Mitigating the Urban Heat Island Effect with an Intensive Green Roof During Summer in Reading, UK.* M.A. Thesis, Reading University.

Boyer, D., and A. Ramaswami. 2017. What Is the Contribution of City-Scale Actions to the Overall Food System's Environmental Impacts? Assessing Water, Greenhouse Gas, and Land Impacts of Future Urban Food Scenarios. *Environmental Science & Technology* 51: 12035–45.

Boylan, T.A. 1997. Book Reviews. *Regional Studies* 31: 631–8.

Broad, G.M. 2016. *More Than Just Food: Food Justice and Community Change*. Berkeley: University of California Press.

Broadstone, S., and C. Brannstrom. 2017. 'Growing Food Is Work': The Labour Challenges of Urban Agriculture in Houston Texas. In, *Global Urban Agriculture*, edited by A.M.G.A. WinklerPrins, 66–78. Boston, MA: CABInternational.

Brown, K.H., and A.L. Jameton. 2000. Public Health Implications of Urban Agriculture. *Journal of Public Health Policy* 21: 20–39.

Cadieux, K.V., and R. Slocum. 2015. What Does It Mean to Do Food Justice? *Journal of Political Ecology* 22: 1–26.

Carlisle, L. 2013. Critical Agrarianism. *Renewable Agriculture and Food Systems* 29: 135–45.

Cassidy, E., P. West, J. Gerber, and J. Foley. 2013. Redefining Agricultural Yields: From Tonnes to People Nourished per Hectare. *Environmental Research Letters* 8: 1–8.

Cattell, V., N. Dines, W. Gesler, and S. Curtis. 2008. Mingling, Observing, and Lingering: Everyday Public Spaces and Their Implications for Well-Being and Social Relations. *Health & Place* 14: 544–61.

CCAFS. 2012. *Is Eating Local Good for the Climate? Thinking Beyond Food Miles*. Wageningen: Climate Change, Agriculture, and Food Security, November 6. https://ccafs.cgiar.org/blog/eating-local-good-climate-thinking-beyond-food miles#.WVwXFojyvan (Accessed 5 August 2020).

CDC. 2010. *Community gardens Resource Library/White Papers*. Atlanta, GA: US Centers for Disease Control and Prevention. www.cdc.gov/ (Accessed 5 August 2020).

CDC. 2020. Certain Medical Conditions and Risks for Severe COVID-19 Illness. www.cdc.gov/coronavirus (Accessed 5 August 2020).

CFF. 2017. *Carbon Footprint Factsheet*. Ann Arbor: Center for Sustainable Systems, University of Michigan. http://css.umich.edu/factsheets/carbon-footprint-factsheet (Accessed 5 August 2020).

Chemisana, D., and C. Lamnatou. 2014. Photovoltaic-Green Roofs: An Experimental Evaluation of System Performance. *Applied Energy* 119: 246–56.

Cheng, Z., A. Paltseva, I. Li, T. Morin, H. Huot, S. Egendorf, Z. Su, R. Yolanda, K. Singh, L. Lee, M. Grinshtein, Y. Liu, K. Green, W. Wai, B. Wazed, and R. Shaw. 2015. Trace Metal Contamination in New York City Garden Soils. *Soil Science* 180(4/5): 1–8.

Christensen, S., P. Malberg Dyg, and K. Allenberg. 2019. Urban Community Gardening, Social Capital, and "Integration"—A Mixed Method Exploration of Urban "Integration-Gardening" in Copenhagen, Denmark. *Local Environment* 24(3): 231–48.

Cohen, N., and K. Reylolds. 2015. Resource Needs for a Socially Just and Sustainable Urban Agriculture System: Lessons from New York City. *Renewable Agriculture and Food Systems* 30: 103–14.

Colasanti, K., C. Litjens, and M. Hamm. 2010. *Growing Food in the City: The Production Potential of Detroit's Vacant Land*. East Lansing, MI: The CS Mott Group for Sustainable Food Systems, Michigan State University.

Comstock, N., M. Dickinson, J. Marshall, M.-J. Soobader, M. Turbin, and M. Buchenau. 2010. Neighborhood Attachment and Its Correlates: Exploring Neighborhood Conditions, Collective Efficacy, and Gardening. *Journal of Environmental Psychology* 30: 435–42.

Conway, T.M. 2016. Home-Based Edible Gardening: Urban Residents' Motivations and Barriers. *Cities and the Environment (CATE)* 9(1): Article 3. https://digitalcommons.lmu.edu/cgi/viewcontent.cgi?article=1203&context=cate (Accessed 13 May 2020).

Coote, A. 2014. *A New Social Settlement for People and Planet: Understanding the Links between Social Justice and Sustainability*. London: New Economics Foundation.

Corcoran, M.P., and J.S. Cavino. 2018. Introduction. In, *Civil Society and Urban Agriculture in Europe*, edited by M.P. Corcoran, and J.S. Cavino, 1–16. Oxford: Berghahn.

Cowley, S. 2015. Near Detroit, Florist's Vision Turns Abandoned House into Art. *The New York Times*, 15 October: B8.

Datta, R. 2019. Sustainability Through Cross-Cultural Community Garden Activities. *Local Environment* 24(8): 762–76.

Davolio, F., and R. Sassatelli. 2009. Foodies Aesthetics and Their Reconciliatory View of Food Politics. *Sociologica* 1: 1–7.

De Bon, H., L. Parrot, and P. Moustier. 2010. Sustainable Urban Agriculture in Developing Countries: A Review. *Agronomy for Sustainable Development* 30: 21–32.

De la Salle, J., and M. Holland. 2010. *Agricultural Urbanism: Handbook for Building Sustainable Food & Agriculture Systems in 21st Century Cities*. Faringdon, Oxfordshire: Green Frigate Books.

De Lind, L. 2011. Are Local Food and the Local Food Movement Taking Us Where We Want to Go? Or Are We Hitching Our Wagons to the Wrong Stars? *Agriculture and Human Values* 28: 273–83.

Desmond, D., J. Grieshop, and A. Subramaniam. 2004. *Revisiting Garden-Based Learning in Basic Education*. Rome: FAO.

Despommier, D. 2009. The Rise of Vertical Farms. *Scientific American* 301: 66–7.

Despommier, D. 2010. *The Vertical Farm: Feeding the World in the 21st Century*. New York: St. Martin's Press.

Despommier, D. 2017. Vertical Farming Using Hydroponics and Aeroponics. In, *Urban Soils*, edited by R. Lal, and B.A. Stewart, 313–28. Boca Raton, FL: CRC Press.

Dubbeling, M., F. Orsini, and G. Giaquinto. 2017. Introduction. In, *Rooftop Urban Agriculture*, edited by F. Orsini, M. Dubbeling, H. de Zeeuw, and G. Giaquinto, 3–8. Berlin: Springer.

Duffy, M. 2009. Economies of Size in Production Agriculture. *Journal of Hunger & Environmental Nutrition* 4: 375–92.

Duggan, Tara. 2018. The Bay Area Company Building World's Largest Vertical farm in Dubai. *San Francisco Chronicle*, 20 July. www.sfchronicle.com/food/article/The-Bay-Area-company-building-world-s-largest-13092905.php (Accessed 5 August 2020).

DuPuis, M., and D. Goodman. 2005. Should We Go 'Home' to Eat: Toward a Reflexive Politics of Localism. *Journal of Rural Studies* 21: 359–71.

DuPuis, M., T. Marsden, and J. Murdoch, eds. 2006. *Between the Local and the Global: Confronting Complexity in the Contemporary Agri-Food Sector*. Bingley: Emerald Group.

Durrant, R. 2016. *Civil Society Roles in Transition: Towards Sustainable Food*? SLRG Working Paper 02–14. Sustainable Lifestyles Research Group, University of Surrey, Guildford.

Edmondson, J.L., H. Cunningham, D.O. Densley Tingley, M.C. Dobson, D.R. Grafius, J.R. Leake, N. McHugh, J. Nickles, G.K. Phoenix, A.J. Ryan, V. Stovin, N. Taylor Buck, P.H. Warren, and D.D. Cameron. 2020. The Hidden Potential of Urban Horticulture. *Nature Food* 1: 155–9.

Eizenberg, E. 2012. Actually Existing Commons: Three Moments of Space of Community Gardens in New York City. *Antipode* 44: 764–82.

Engel-Di Mauro, S., and C. Cattaneo. 2014. Squats in Urban Ecosystems: Overcoming the Social and Ecological Catastrophes of the Capitalist City. In, *The Squatters' Movement in Europe: Commons and Autonomy as Alternatives to Capitalism*, edited by C. Cattaneo, and M. Martínez, 166–88. London: Pluto.

FAO. 2010. *Growing Greener Cities*. Rome: Food and Agriculture Organization, United Nations. www.fao.org/ag/agp/greenercities/pdf/GGC-en.pdf (Accessed 14 January 2017).

FAO. 2013. *Statistical Yearbook*. Rome: Food and Agriculture Organization, United Nations.

FAO. 2017. *The State of Food Security and Nutrition in the World 2017: Building Resilience for Peace and Food Security*. Rome: Food and Agriculture Organisation, United Nations.

FCRN. 2012. *Food Printing Oxford: How to Feed a City*. Oxford: Food and Climate Research Network.

FCRN. 2015. *The Environmental Impacts of Food: An Introduction to LCA*. Oxford: Food and Climate Research Network. https://foodsource.org.uk/24-value-and-limitations-lifecycle-assessment (Accessed 15 June 2020).

Ferris, J., C. Norman, and J. Sempik. 2001. People, Land and Sustainability: Community Gardens and the Social Dimension of Sustainable Development. *Social Policy & Administration* 35: 559–68.

Firth, C., D. Maye, and D. Pearson. 2011. Developing 'Community' in Community Gardens. *Local Environment* 16: 555–68.

Fisher, S., and A. Karunanithi. 2014. *Contemporary Comparative LCA of Commercial Farming and Urban Agriculture for Selected Fresh Vegetables Consumed in Denver, Colorado*. San Francisco, CA: LCA Food Conference.

Foderaro, L.W. 2012. In the Book Bag, more Garden Tools. *The New York Times*, 24 November: A16.

fresh. 2012. *Edmonton's Food & Urban Agriculture Strategy*. Edmonton AB: City of Edmonton.

Freund, P. 2008. The Expressive Body: A Common Ground for the Sociology of Emotions and Health and Illness. *Sociology of Health & Illness* 12: 452–77.

Freund, P., and G. Martin. 2008. Fast Cars/Fast Food: Hyperconsumption and Its Health and Environmental Consequences. *Social Theory & Health* 6: 309–22.

Fuller, T. 2016. Healthy Food Everywhere but on the Table. *The New York Times*, 16 November: A19, A25.

Garnett, T. 2000. *Urban Agriculture in London: Rethinking Our Food Economy*. Leiden: The RUAF Foundation. www.ruaf.org/sites/default/files/London_1.pdf (Accessed 5 July 2018).

Giddings, B., B. Hopwood, and G. O'Brien. 2002. Environment, Economy and Society: Fitting Them Together into Sustainable Development. *Sustainable Development* 10: 187–96.

Gliessman, S.R. 2016. Agroecology: Roots of Resistance to Industrial Food Systems. In, *Agroecology. A Transdisciplinary, Participatory and Action-Oriented Approach*, edited by V.E. Méndez, C.M. Bacon, R. Cohen, and S.R. Gliessman. Boca Raton, FL: CRC Press.

Golden, S. 2013. *Urban Agriculture Impacts, Social, Health, and Economic: A Literature Review*. Davis: Agricultural Sustainability Institute, University of California.

Graefe, D., A. Mowen, and A. Graefe. 2017. Craft Beer Enthusiasts' Support for Neolocalism and Environmental Causes. *Craft Beverages and Tourism* 2: 27–47.

Gray, L., L. Diekmann, and S. Algert. 2017. North American Urban Agriculture: Barriers and Benefits. In, *Global Urban Agriculture*, edited by A.M.G.A. WinklerPrins, 24–37. Boston, MA: CABInternational.

Grewal, S., and P. Grewal. 2012. Can Cities Become Self-Reliant in Food? *Cities* 29: 1–11.

Grubinger, V. 2010. *Ten Reasons to Buy Local Food*. Burlington: Vermont Vegetable and Berry Program, University of Vermont Extension. www.uv.edu/vtveganandberry/factssheets/buylocal.html (Accessed 6 July 2018).

Gunia, A. 2020. Why Coronavirus Could Cause Millions of People to Go Hungry, Even Though There's Enough Food to Go Around. *Time Magazine*, 8 May. https://time.com/5820381/coronavirus-food-shortages-hunger/ (Accessed 10 May 2020).

Hamm, M.W. 2016. *Feeding Cities—With Indoor Vertical Farms? City Farmer*. East Lansing: Michigan State University, Center for Regional Food Systems. www.cityfarmer.info/2016/03/26/ (Accessed 8 July 2018).

Harada, Y., T.H. Whitlow, J. Russell-Anelli, M.T. Walter, N.L. Bassuk, and M.A. Rutzke. 2019. The Heavy Metal Budget of an Urban Rooftop Farm. *Science of The Total Environment* 660: 115–25.

Harris, P. 2010. Detroit Gets Growing. *The Observer Magazine* 11: 42–9.

Hein, J.R., B. Ilbery, and M. Knefsey. 2006. Distribution of Local Food Activity in England and Wales: An Index of Food Relocalization. *Regional Studies* 40: 289–301.

Hess, D.J. 2009. *Localist Movements in a Global Economy: Sustainability, Justice, and Urban Development in the United States*. Cambridge, MA: The MIT Press.

Hickey, S. 2015. Tunnel Vision Turns Air-Raid Shelter into Hi-Tech Farm Supplying Herbs to Top London Restaurants. *The Guardian*, 14 September 14: 27.

Hirsch, F. 2005. *Social Limits to Growth*. London: Routledge.

Holland, L. 2004. Diversity and Connections in Community Gardens: A Connection to Local Sustainability. *Local Environment* 9: 285–305.

Holt-Giménez, E. 2017. *A Foodie's Guide to Capitalism: Understanding the Political Economy of What We Eat*. New York: NYU Press.

Holt-Gimenez, E., and A. Shattuck. 2011. Food Crises, Food Regimes and Food Movements: Rumblings of Reform or Tides of Transformation? *The Journal of Peasant Studies* 38(1): 109–44.

Holtkamp, C., T. Shelton, G. Daly, C.C. Hiner, and R.R. Hagelman III. 2016. Assessing Neolocalism in Microbreweries. *Papers in Applied Geography* 2: 66–78.

InterAcademy Partnership. 2018. Opportunities for Future Research and Innovation on Food and Nutrition Security and Agriculture. In, *The InterAcademy Partnership's Global Perspective*. Washington, DC: Inter Academy Partnership. www.knaw.nl/shared/resources/internationaal/bestanden/IAPFNSAWebcomplete16Nov2018.pdf (Accessed 13 June 2019).

IPCC. 2014. Report of Working Group II: Impacts, Adaptation and Vulnerability. *Fifth Assessment Report*. www.ipcc.ch/report/ar5/wg2 (Accessed 10 July 2017).

Irvine, S., L. Johnson, and K. Peters. 1999. Community Gardens and Sustainable Land Use Planning: A Case-Study of the Alex Wilson Community Garden. *Local Environment* 4: 33–46.

Jackson, T. 2009. *Prosperity Without Growth: Economics for a Finite Planet*. London: Routledge.

Jácome-Pólit, D., D. Paredes, A. Santandreu, A. Rodríguez Dueñas, and N. Pinto. 2019. Quito's Resilient Agrifood System. *ISOCARP Review* 15: 276–300. https://ruaf.org/assets/2020/01/Quitos-Resilient-Agrifood-System-1.pdf (Accessed 14 July 2020).

Johnson, R., R.A. Aussenberg, and T. Cowan. 2013. *The Role of Local Food Systems in U.S. Farm Policy*. Washington, DC: US Congressional Research Service. www.nardep. info/uploads (Accessed 5 August 2020).

Johnston, J., and S. Baumann. 2009. Tension in the Kitchen: Explicit and Implicit Politics in the Gourmet Foodscape. *Sociologica* 1. www.sociologica.mulino.it/doi/ 10.2383/29565.

Karg, H., P. Drechsel, E.K. Akoto-Danso, R. Glaser, G. Nyarko, and A. Buerkert. 2016. Foodsheds and City Region Food Systems in Two West African Cities. *Sustainability* 8: 1175.

Kingsley, J., E. Foenander, and A. Bailey. 2019. 'You Feel Like You're Part of Something Bigger': Exploring Motivations for Community Garden Participation in Melbourne, Australia. *BMC Public Health* 19: 745.

Kinnunen, P., et al. 2020. Local Food Crop Production Can Fulfil Demand for Less Than One-Third of the Population. *Nature Food* 1: 229–37.

Kline, C., S.L. Slocum, and C.T. Cavaliere. 2017. The Impact and Implications of Craft Beer Research: An Interdisciplinary Literature Review. In, *Craft Beverages and Tourism,*

Volume 1: The Rise of Breweries and Distilleries in the United States, edited by C. Kline, S.L. Slocum, and C.T. Cavaliere, 11–24. New York: Springer.

Klink, J., et al. 2011. Metropolitan Fragmentation and Neo-localism in the Periphery: Revisiting the Case of Curitiba. *Urban Studies* 49(3): 543–61.

Kuhnen, M., and R. Hahn. 2017. Indicators in Social Life Cycle Assessment: A Review of Frameworks, Theories, and Empirical Experience. *Journal of Industrial Ecology* 21(6). doi: 10.1111/jiec.12663.

Kulak, M., A. Graves, and J. Chatterton. 2013. Reducing Greenhouse Gas Emissions with Urban Agriculture: A Life Cycle Assessment Perspective. *Landscape and Urban Planning* 111: 68–78.

Kummu, M., H. de Moel, M. Porkka, S. Siebert, O. Varis, and P. Ward. 2012. Lost Food, Wasted Resources: Global Food Supply Chain Losses and Their Impacts on Freshwater, Cropland, and Fertilizer Use. *Science of the Total Environment* 438: 477–89.

Ladner, P. 2011. *The Urban Food Revolution*. Gabriola Island BC: New Society.

Langellotto, G.A., A. Melathopoulos, I. Messer, A. Anderson, N. McClintock, and L. Costner. 2018. Garden Pollinators and the Potential for Ecosystem Service Flow to Urban and Peri-Urban Agriculture. *Sustainability* 10: 2047. doi: 10.3390/su10062047.

Lawrance, B., and C. de la Pena. 2012. Introduction. In, *Local Foods Meet Global Foodways: Tasting History*, edited by B. Lawrance, and C. Pena, 2–14. Abingdon: Taylor & Francis.

Leahy, T. 2018. Radical Reformism and the Marxist Critique. *Capitalism Nature Socialism* 29: 61–74.

Lefebvre, H. 2000 [1974]. *The Production of Space*. Translated by D. Nicolson-Smith. Oxford: Blackwell.

Li, D., E. Bou-Zeid, and M. Oppenheimer. 2014. The Effectiveness of Cool and Green Roofs as Urban Heat Island Mitigation Strategies. *Environmental Research Letters* 9: 055002.

Li, I., Z. Cheng, A. Paltseva, T. Morin, B. Smith, and R. Shaw. 2017. Lead in New York City Soils. In, *Megacities 2050: Environmental Consequences of Urbanization*, edited by V.I. Vasenev, E. Dovletyarova, Z. Cheng, and R. Valentini, 62–79. Berlin: Springer Geography.

Li, Y., and R.W. Babcock. 2014. Green Roofs against Pollution and Climate Change: A Review. *Agronomy for Sustainable Development* 34: 695–705.

Litt, J.S., et al. 2011. The Influence of Social Involvement, Neighborhood Aesthetics, and Community Garden Participation on Fruit and Vegetable Consumption. *American Journal of Public Health* 101: 1466–73.

Litzky, B.E., L. Andersson, and W.P. Smith. 2017. Local Entrepreneurs: Conduits of Neo-Localism and Sustainable Urban Livelihoods. *Academy of Management Proceedings* 2017: 1. https://journals.aom.org/doi/abs/10.5465/ambpp.2017.1096 (Accessed 5 July 2018).

Louv, R. 2008. *Last Child in the Woods: Saving Our Children from Nature-Deficit Disorder*. Chapel Hill NC: Algonquin Books.

Lyons, K., C. Richards, L. Desfours, and M. Amati. 2013. Food in the City: Urban Food Movements and the (Re)-Imagining of Urban Spaces. *Australian Planner* 50: 157–63.

Lyson, T.A. 2004. *Civic Agriculture: Reconnecting Farm, Food, and Community*. Medford, MA: Tufts University Press.

Macdiarmid, J.I., J. Kyle, G.W. Horgan, J. Loe, C. Fyfe, A. Johnstone, and G. McNeill. 2012. Sustainable Diets for the Future: Can We Contribute to Reducing Greenhouse Gas Emissions by Eating a Healthy Diet? *American Journal of Clinical Nutrition* 96: 632–9.

Mandel, L. 2013. *Eat Up: The Inside Scoop on Rooftop Agriculture*. Gabriola Island BC: New Society Publishers.

Mares, T.M., and D.G. Pena. 2010. Urban Agriculture in the Making of Insurgent Spaces in Los Angeles and Seattle. In, *Guerrilla Urbanism and the Remaking of Contemporary Cities*, edited by J. Hou, 241–54. London: Routledge.
Mark, J. 2020. The Rebirth of the Food Sovereignty Movement. www.sierraclub.org/sierra/rebirth-food-sovereignty-movement (Accessed 20 April 2020).
Marsden, J. 2015. An Economic Analysis of London's Housing Market. *GLA Economics*, 15 December. www.london.gov.uk/sites/default/files/151215_joel_marsde_house_prices_in_london.pdf (Accessed 12 August 2018).
Martellozzo, F., J.-S. Landry, D. Plouffe, V. Seufert, P. Rowhani, and N. Ramankutty. 2015. Urban Agriculture: A Global Analysis of the Space Constraint to Meet Urban Vegetable demand. *Environmental Research Letters* 9: 064025. http://iopscience.iop.org/1748–9326/9/6/064025/ (Accessed 25 March 2016).
Martin, C. 2016. Taking Local Produce to Another Level. *The New York Times*, 27 March: BU4.
Martin, G. 2013. *Urban Agriculture's Synergies with Ecological and Social Sustainability: Food, Nature, and Community*. Brighton: European Conference on Sustainability, Energy & the Environment.
Martin, G., R. Clift, and I. Christie. 2016. Urban Cultivation and Its Contributions to Sustainability: Nibbles of Food but Oodles of Social Capital. *Sustainability* 8(409): 1–18.
McClintock, N. 2010. Why Farm the City? Theorizing Urban Agriculture through a Lens of Metabolic Rift. *Cambridge Journal of Regions, Economy and Society* 3: 191–207.
McClintock, N. 2011. From Industrial Garden to Food Desert: Demarcated Devaluation in the Flatlands of Oakland, California. In, *Cultivating Food Justice: Race, Class, and Sustainability*, edited by A.H. Alkon, and J. Agyeman, 89–120. Cambridge MA: MIT Press.
McClintock, N. 2014. Radical, Reformist, and Garden-Variety Neoliberal: Coming to Terms with Urban Agriculture's Contradictions. *Local Environment* 19: 147–71.
McClintock, N. 2015. A Critical Physical Geography of Urban Soil Contamination. *Geoforum* 65: 69–85.
McClintock, N., and J. Cooper. 2009. *Cultivating the Commons: An Assessment of the Potential for Urban Agriculture in Oakland's Public Lands*. Oakland: HOPE Collaborative/City Slicker Farms/Food First.
McClintock, N., J. Cooper, and S. Khandeshi. 2013. Assessing the Potential Contribution of Vacant Land to Urban Vegetable Production and Consumption in Oakland, California. *Landscape and Urban Planning* 111: 46–58.
McIlvaine, H., R. Porter, and G. Delany-Barmann. 2019. Change the Game, Not the Rules: The Role of Community Gardens in Disaster Resilience. *Journal of Park and Recreation Administration*. doi: 10.18666/JPRA-2019-9721.
McWilliams, J. 2009. Just Food: Where Locavores Get It Wrong and How We Can Truly Eat Responsibly. *Wall Street Journal*, 22 August. www.wsj.com (Accessed 17 November 2017).
Meemken, E.M., and M. Qaim. 2018. Organic Agriculture, Food Security, and the Environment. *Annual Review of Resource Economics*. www.annualreviews.org/doi/pdf/10.1146/annurev-resource-100517–023252 (Accessed 1 August 2018).
Mees, H., P.P.J. Driessen, H.A.C. Runhaar, and J. Stamatelos. 2013. Who Governs Climate Adaptation? Getting Green Roofs for Stormwater Retention off the Ground. *Journal of Environmental Planning and Management* 56: 802–25.
Meyer, N., and A. Reguant-Closa. 2017. 'Eat as If You Could Save the Planet and Win!': Sustainability Integration into Nutrition for Exercise and Sport. *Nutrients* 9: 412. www.mdpi.com/journal/nutrients (Accessed 14 July 2018).

Mogk, J.E., S. Wiatkowski, and M.J. Weindorf. 2010. Promoting Urban Agriculture as an Alternative Land Use for Vacant Properties in the City of Detroit. *Wayne Law Review* 56: 521–80.

Mohareb, E., M. Heller, P. Novak, B. Goldstein, X. Fonoll, and L. Raskin. 2017. Considerations for Reducing Food System Energy Demand While Scaling Up Urban Agriculture. *Environmental Research Letters* 12: 125004. https://doi.org/10.1088/1748-9326/aa*889b* (Accessed 1 July 2018).

Mok, H.-F., V.G. Williamson, J.R. Grove, K. Burry, S.F. Barker, and A.J. Hamilton. 2014. Strawberry Fields Forever? Urban Agriculture in Developed Countries: A Review. *Agronomy for Sustainable Development* 34: 21–43.

Morgan, K. 2009. Feeding the City: The Challenge of Urban Food Planning. *International Planning Studies* 14: 341–8.

Morgan, K., T. Marsden, and J. Murdoch. 2006. *Worlds of Food: Place, Power and Provenance in the Food Chain*. Oxford: Oxford University Press.

Mougeot, L.J.A. 2006. *Growing Better Cities: Urban Agriculture for Sustainable Development*. Ottawa: IDRC Books.

Muller, A., C. Schader, N. El-Hage Scialabba, J. Brüggemann, A. Isensee, K.-H. Erb, P. Smith, P. Klocke, F. Leiber, M. Stolze, and U. Niggli. 2017. Strategies for Feeding the World More Sustainably with Organic Agriculture. *Nature Communications* 8: Article 1290.

Murray, D.W., and M.A. O'Neill. 2012. Craft Beer: Penetrating a Niche Market. *British Food Journal* 114: 899–909.

Nasr, J., J. Komisar, and H. de Zeeuw. 2017. Panorama of Rooftop Agriculture Types. In, *Rooftop Urban Agriculture*, edited by F. Orsini, M. Dubbeling, H. de Zeeuw, and G. Giaquinto, 9–30. Berlin: Springer.

Nels, T. 2017. *Green Infrastructure: The Need for Soils in Sustainable Cities*. Paper presented at 2nd Annual Urban Soils Symposium, Soils: Our Resource & Our Future. New York City Botanical Garden, November 29–30.

O'Neill, C., D. Houtman, and S. Aupers. 2014. Advertising Real Beer Authenticity Claims Beyond Truth and Falsity. *European Journal of Cultural Studies* 17: 585–601.

Orsini, F., R. Kahane, R. Nono-Womdin, and G. Giaquinto. 2013. Urban Agriculture in the Developing World. *Agronomy for Sustainable Development* 33: 695–720.

Palmer, L. 2018. Urban Agriculture Growth in US Cities. *Nature Sustainability* 1: 5–7.

Parece, T.E., and J.B. Campbell. 2017. A Survey of Urban Community Gardeners in the USA. In, *Global Urban Agriculture*, edited by A.M.G.A. WinklerPrins, 38–49. Boston, MA: CABInternational.

Parnwell, M.J.G. 2007. Neolocalism and Renascent Social Capital in Northeast Thailand. *Environment and Planning D: Society and Space* 25(6): 990–1014.

Payen, S., C. Basset-Mens, and S. Perret. 2015. LCA of Local and Imported Tomato: An Energy and Water Trade-Off. *Journal of Cleaner Production* 87: 139–48.

Perez-Vazquez, A., S. Anderson, and A.W. Rogers. 2005. Assessing Benefits from Allotments as a Component of Urban Agriculture in England. In, *Agropolis: The Social, Political and Environmental Dimensions of Urban Agriculture*, edited by L. Mougeot, 239–66. London: Earthscan.

Peters, C.J., N.L. Bills, J.L. Wilkins, and G.W. Fick. 2008. Foodshed Analysis and Its Relevance to Sustainability. *Renewable Agriculture and Food Systems* 24: 1–7.

Pole, T., and G. Martin. 2017. *What's Brewing? The Role of Craft Beer in Local Communities*. Los Angeles: Agriculture, Food, and Human Values Society & Association for the Study of Food and Society Annual Meeting.

Poulsen, M.N., et al. 2014. Growing an Urban Oasis: A Qualitative Study of the Perceived Benefits of Community Gardening in Baltimore, Maryland. *The Journal of Culture and Agriculture* 36: 69–82.

Poulsen, M.N., R.A. Neff, and P.J. Winch. 2017. The Multifunctionality of Urban Farming: Perceived Benefits for Neighbourhood Improvement. *Local Environment* 22(11): 1411–27.

Pugh, R. 2013. Gardening Is Helping People with Dementia. *The Guardian*, 30 July: 36.

Raymond, C.M., A.P. Diduck, A. Buijs, M. Boerchers, and R. Moquin. 2019. Exploring the Co-Benefits (and Costs) of Home Gardening for Biodiversity Conservation. *Local Environment* 24(3): 258–73.

Razzaghmanesh, M., S. Beecham, and F. Kazemi. 2014. The Growth and Survival of Plants in Urban Green roofs in a Dry Climate. *Science of the Total Environment* 476–7: 288–97.

RCEP. 2007. *The Urban Environment. Royal Commission on Environmental Pollution*, 26th Report. London: The Stationery Office.

Reganold, J.P., and J.M. Wachter. 2016. Organic Agriculture in the Twenty-First Century. *Nature Plants* 2: Article 15221.

Relf, D., ed. 1992. *The Role of Horticulture in Human Well-Being and Social Development*. Portland OR: Timber Press.

Rock, M., S. Engel-Di Mauro, S. Chen, M. Iachetta, A. Mabey, K. McGill, and J. Zhao. 2017. Food Production in Chongqing, China: Opportunities and Challenges. *Middle States Geographer* 49: 55–62.

Rosin, C., P. Stock, and H. Campbell, eds. 2012. *Food Systems Failure: The Global Food Crisis and the Future of Agriculture*. London and New York: Earthscan.

Rumble, H., and A.C. Gange. 2013. Soil Microarthropod Community Dynamics in Extensive Green Roofs. *Ecological Engineering* 57: 197–204.

Rusciano, V., G. Civero, and D. Scarpato. 2017. Urban Gardening as a New Frontier of Wellness: Case Studies from the City of Naples. *The International Journal of Sustainability in Economic, Social, and Cultural Context* 13: 39–49.

Santo, R., A. Palmer, and B. Kim. 2016. *Vacant Lots to Vibrant Plots: A Review of the Benefits and Limitations of Urban Agriculture*. Baltimore, MD: Center for a Livable Future, Johns Hopkin University.

Sanye-Mengual, E., J. Olive-Sola, A. Anton, J.I. Montero, and J. Rieradevail. 2014. *Environmental Assessment of Urban Horticulture Structures: Implementing Rooftop Greenhouses in Mediterranean Cities*. San Francisco, CA: Life Cycle Assessment—Food Conference.

Scelzi, R., and L. La Fortuna. 2015. Social Food: A Semioethic Perspective on Foodism and New Media. *Southern Semiotic Review* 5: 127–40.

Schlosser, E. 2002. *Fast Food Nation: The Dark Side of the All-American Meal*. New York: HarperCollins.

Schnell, S.M., and J.F. Reese. 2013. Deliberate Identities: Becoming Local in a Global Age. *Journal of Cultural Geography* 30: 55–89.

Schnell, S.M., and J.F. Reese. 2014. Microbreweries, Place, and Identity in the United States. In, *The Geography of Beer*, edited by M. Patterson, and N. Hoalst-Pullen, 167–87. Dordrecht: Springer.

Schwartzman, P., and D. Schwartzman. 2019. *The Earth Is Not for Sale: A Path Out of Fossil Capitalism to the Other World That is Still Possible*. New York: World Scientific.

Schweitzer, O., and E. Erell. 2014. Evaluation of the Energy Performance and Irrigation Requirements of Extensive Green Roofs in a Water-Scarce Mediterranean Climate. *Energy and Buildings* 68: 25–32.

Scott, A., A. Dean, V. Barry, and R. Kotter. 2017. Places of Urban Disorder? Exposing the Hidden Nature and Values of an English Private Urban Allotment Landscape. *Landscape and Urban Planning* 169: 185–96.

SDC. 2008. *Health, Place and Nature*. London: Sustainable Development Commission.

Seidl, M., M.C. Gromaire, M. Saad, and B. De Gouvello. 2013. Effect of Substrate Depth and Rain-Event History on the Pollutant Abatement of Green Roofs. *Environmental Pollution* 183: 195–203.

Silva, J. 2016. *More Than Just Food: Considering the Links between Small-Scale Food and Agriculture, Community Resilience and Food Nutrition Security in Victoria, Canada*. Aarhus: Master of Science, Aarhus University.

Singleton, C.R., B. Sen, and O. Affuso. 2015. Disparities in the Availability of Farmers Markets in the United States. *Environmental Justice* 8: 135–43.

Smith, C. 2010. *London: Garden City*? London: London Wildlife Trust.

Smith, N. 2008 [1984]. *Uneven Development: Nature, Capital, and the Production of Space*. Athens: University of Georgia Press.

Sonnino, R. 2013. Local Foodscapes: Place and Power in the Agri-Food System. *Acta Acriculturae Scandinavica, Section B: Soil & Plant Science* 63: 2–7.

Southard, D. 2014. *Rethinking Localism in Sustainable Development Practice: The Case of Development Monks in Northeast Thailand*. Khon Kaen: The 4th Khon Kaen University National and International Conference.

Specht, K., et al. 2014. Urban Agriculture of the Future: An Overview of Sustainability Aspects of Food Production in and on Buildings. *Agriculture and Human Values* 31: 33–51.

Stancu, C., and A. Smith. 2006. *Food Miles—The International Debate and Implications for New Zealand Exporters*. Auckland: Sustainability & Society, Landcare Research. https:www.landcareresearch.co.nz/food_miles.pdf (Accessed 9 July 2018).

Starrs, P.F., and P. Goin. 2010. *Field Guide to California Agriculture*. Berkeley: University of California Press.

Steier, G. 2018. *Advancing Food Integrity: GMO Regulation, Agroecology, and Urban Agriculture*. Boca Raton, FL: CRC Press.

Subramani, T., and R. Selvan. 2014. Developing a Planning Framework for Accessible and Sustained Urban Agriculture. *International Journal of Engineering Research and Applications* 4: 180–90.

Taylor, J.R., and S. Lovell. 2012. Mapping Public and Private Spaces of Urban Agriculture in Chicago through Analysis of High-Resolution Aerial Images in Google Earth. *Landscape and Urban Planning* 108: 57.

Thebo, A.L., et al. 2014. Global Assessment of Urban and Peri-Urban Agriculture: Irrigation and Rainfed Croplands. *Environmental Research Letters* 9: 114002.

Thuring, C.E., and N. Dunnett. 2014. Vegetable Composition of Old Extensive Green Roofs (from 1980s Germany). *Ecological Processes* 3: 4.

Turner, B. 2011. Embodied Connections: Sustainability, Food Systems and Community Gardens. *Local Environment* 16: 509–22.

UN. 1987. *Report of the World Commission on Environment and Development*. New York: United Nations 96th Plenary Meeting, 11 December. www.un.org/documents/ga/res/42/ares42–187.htm (Accessed 18 July 2018).

USDA. 2013. *Climate Change and Agriculture in the United States: Effects and Adaptation*. Technical Bulletin 1935. Washington, DC: US Department of Agriculture.

USDA. 2014. *2012 Census of Agriculture: U.S. Farms and Farmers*. Washington, DC: US Department of Agriculture.

Vallance, S., H.C. Perkins, and J.E. Dixon. 2011. What Is Social Sustainability? A Clarification of Concepts. *Geoforum* 42: 342–8.

Vijayaraghavan, K. 2016. Green Roofs: A Critical Review on the Role of Components, Benefits, Limitations and Trends. *Renewable and Sustainable Energy Reviews* 57: 740–52.

Vijayaraghavan, K., and U.M. Joshi. 2014. Can Green Roof Act as Sink for Contaminants? A Methodological Study to Evaluate Runoff Quality from Roofs. *Environmental Pollution* 195: 121–9.

Viljoen, A. 2005. *Continuous Productive Urban Landscapes: Designing Urban Agriculture for Sustainable Cities*. London: Architectural Press.

Vitiello, D., and M. Nairn. 2009. *Community Gardening in Philadelphia*. Philadelphia, PA: Penn Planning and Urban Studies, University of Pennsylvania. www.farmlandinfo.org/sites/default/files/Pgladelphia_Harvest_1.pdf (Accessed 5 August 2020).

Wallace, R. 2016. *Big Farms Make Big Flu: Dispatches on Infectious Disease, Agribusiness, and the Nature of Science*. New York: Monthly Review Press.

Waters, Alice. 2011. *40 Years of Chez Panisse*. New York: Random House.

Wegerif, M.C.A., and J.S.C. Wiskerke. 2017. Exploring the Staple Foodscape of Dar es Salaam. *Sustainability* 9: 1081.

Weinstein, N., A. Balmford, C.R. Dehaan, V. Gladwell, R.B. Bradbury, and T. Amano. 2015. Seeing Community for the Trees: Links Between Contact with Natural Environments, Community Cohesion, and Crime. *BioScience* 65: 1141–53.

Weissman, E. 2014. Brooklyn's Agrarian Questions. *Renewable Agriculture and Food Systems* 30: 92–102.

White, R., and A. Sterling. 2013. Sustaining Trajectories towards Sustainability: Dynamics and Diversity in UK Communal Growing Activities. *Global Environmental Change* 23: 838–46.

Whittinghill, L.J., D.B. Rowe, and B.M. Cregg. 2013. Evaluation of Vegetable Production on Extensive Green Roofs. *Agroecology and Sustainable Food Systems* 37: 465–84.

WHO. 2000. *CINDI Dietary Guide*. Copenhagen: World Health Organization Office for Europe.

Wilson, E.O. 1984. *Biophilia*. Cambridge MA: Harvard University Press.

WinklerPrins, A.M.G.A. 2017. Defining and Theorizing Global Urban Agriculture. In, *Global Urban Agriculture*, edited by A.M.G.A WinklerPrins, 1–11. Boston, MA: CABInternational.

Winnie, H. 2015. Hot Peppers Becoming a Cash Crop for Bronx Community Garden. *The New York Times*, 19 June: A15.

Wolch, J.R., J. Byrne, and J.P. Newell. 2014. Urban Green Space, Public Health, and Environmental Justice: The Challenge of Making Cities 'Just Green Enough'. *Landscape and Urban Planning* 125: 234–44.

Yuan, L. 2015. Could Underground Farms Be the Future of Urban Agriculture: The Lowline Lab Gives Us a Taste of Subterranean Greens. *Mold*, 11 June. https://thisismold.com/ space/farm-systemscould-underground/ (Accessed 9 July 2018).

Zoll, F., K. Specht, I. Opitz, R. Siebert, and A. Piorr. 2018. Individual Choice or Collective Action? Exploring Consumer Motives for Participating in Alternative Food Networks. *International Journal of Consumer Studies* 42: 101–10.

6

ECOSOCIAL CHALLENGES AND IMPACTS OF URBAN FOOD PRODUCTION

As shown in the previous chapter, because of its limited food-producing possibilities but much more promising social effects, urban agriculture is more usefully understood as urban cultivation, especially in the sense of contributing to social capital and the development of human potentials. Broaching issues of power dynamics is essential in all this. But knowing only about the social relations fostered through urban cultivation unnecessarily constrains one's grasp of the overall effects of urban food production as well as its political potentials. Moreover, city food presents many kinds of situations that impel a careful and comparative study of localities so that unsupported generalisation is avoided. As discussed in Chapter 4, the biophysical dimensions of cities must also be considered. This is necessary even as the city is a site of particularly heavy-handed human influence that alters the physical environment in stratigraphically lasting ways and modifies the composition of and relations among species, with beneficial effects for some and detrimental ones for others (Botkin and Beveridge 1997; Childers et al. 2015; Clucas and Marzluff 2011). Urban food production is a form among others of ecological transformation as much as it is also a set of biophysically constrained practices. It affects and is affected by other human and non-human activities. To actualise and successfully spread urban cultivation, there are both biophysical and social contingencies to be considered that may be constraining or enhancing.

Giving prominence to biophysical factors in this chapter does not mean losing track of social power relations. Far from it! An ecosocial framework is underpinned by historical and dialectical materialism, where physical (environmental), ecological, and social factors are studied as mutually influential but not equivalent processes (on this, see the exemplary work of Levins and Lewontin 1985; see also Harvey 1996). Depending on place (i.e. variable configurations of factors) and physical scale (i.e. how large an area), some factors are more influential than others. Social relations, in our case, are more decisive in urban areas, as is urban

DOI: 10.4324/9781003131281-6

cultivation in particular, but they are dialectically related to physical forces (and the environments thereby produced) and the ecosystems of which they are a constitutive part. By dialectical we mean in the sense that biophysical dynamics and social relations affect and transform each other, and that the resulting transformation involves people changing themselves as they bring about changes in ecosystems and environments. This is the basis of what we mean by ecosocial relations. But this mutual transformative character in ecosocial relations is never on an equal footing. Social relations are one constituent of urban ecosystems among many, and they are certainly never the sole causal process. Some refer to this as recognising the "more-than-human" aspect of the city (or whatever else), but that kind of wording seems to us redundant with what ecologists have already come up with for a long time.

Some of these considerations are rather banal. It turns out that most of the biophysical factors involved in urban food production are generally the same as for any farming system. Growing crops (including fruit trees, herbs, mushrooms, etc.) requires accessibility to land, adequate sunlight, water, and nutrients (including attention to amounts and micronutrient types relative to species), as well as proper temperature ranges, promotion of symbioses, and effective defence from pathogens. Raising land-based animals (e.g. alpacas, chickens, cows, geese, goats, rabbits, and water buffalo) implies provision of water and species-specific food and area, defence from predators and parasites, encouragement of mutualistic relations, and pasture for large herbivores. Amphibians and fish necessitate sufficiently deep, well-oxygenated, relatively clear freshwater (or brackish or sea water for crustaceans, other fish). These are a few examples of what to consider, and in many farming communities, especially those that are subsistence-oriented, there is plenty of expertise to overcome the challenges. The feasibility of food production depends on successfully attuning species selection and requirements to local environmental and ecological conditions.

Generally, a great deal of knowledge and experience is usually needed to be able to produce food. Cities where peasants, gatherer-hunters, and fishers have migrated can greatly benefit from the availability of such expertise in the establishment and expansion of urban cultivation areas, provided their knowledge is not ignored. People from largely self-reliant, subsistence-oriented communities also tend to be highly skilled and innovative in food procurement and production and in finding ways to overcome biophysical challenges, as attested in many studies (Altieri 1987; Brookfield 2001; Posey and Balée 1989; Richards 1985). Still, there are biophysical factors in cities often induced by enduring industrial impacts that diverge from those encountered in other ecosystems (this is the main reason for the lengthy discussion in Chapter 4 on the biophysical peculiarities of cities). To some extent, such lasting impacts as found in cities can also be found in other areas (like mines and rural industrial complexes) and may be much more intensely destructive there (e.g. deforestation, river pollution). So, some of the problems will be similar and entail similar kinds of attention and solutions. In what follows, we give an overview of the major factors involved that overlap with those obtained in other ecosystems yet exhibit characteristics specific to urban situations.

Biophysical factors affecting and affected by urban cultivation

Climate type exerts a prominent influence in terms of the amount and timing of sunlight and precipitation and temperature ranges and oscillations, all of which affect the ability of producing crops. Cities in arid or semi-arid regions should beckon careful selection of drought-tolerant species, if water use and groundwater withdrawals are to be environmentally sustainable. Location within climates of high precipitation and sunlight affords year-round productivity that cannot be achieved in more poleward latitudes without recourse to energy-demanding greenhouses. In subarctic towns, up to 24-hour daylight time greatly enhances vegetable production for several months. Even if weather conditions tend to be too harsh for plant growth for most of the year, growing seasons have been steadily expanding with global warming so that urban cultivation may become ironically more feasible at such latitudes (Barbeau et al. 2018), while contributing, if carried out alongside various current forms of colonialism, to the erosion of Indigenous peoples' livelihoods and autonomous food procurement strategies.

More localised human-induced climatic shifts also have lasting repercussions on urban cultivation potentials. Urban heat island effects enable the growing of vegetables or fruit trees that otherwise would not survive local climate conditions, like figs and rosemary grown in parts of New York City. At the same time, pathogens and unwanted plants ("weeds") can also proliferate with warmer conditions, and plants that require cooler temperatures will flower early, or grow poorly if at all. Urban heat island effects may accentuate the magnitude of heat waves, which can impair plant growth (e.g. rapid water loss and wilting) and endanger human and many other animals' health. Higher overall average summer-time temperatures in cities, especially in urban areas with sparse vegetation cover, can make food production difficult by restricting the time when it is safer to tend to vegetables.

Landform position affects the intensity of sunlight received even without the presence of large, tall buildings. Establishing cultivated areas on a leeward or southern-facing side of steep hilly areas (over 500 metres) can also result in appreciably different amounts of sunlight (and possibly precipitation), with consequently greater amounts of irrigation water needed. Cultivated spaces in low-lying areas may have ease of groundwater availability, which is important in places devoid of piping systems. But they may also be more flood-prone or, in coastal zones, susceptible to sea-water inundation, especially during severe weather events. Sea-water intrusion, increasingly prominent in places where sea-level rise combines with high groundwater withdrawal, can also reduce irrigation water availability. Contaminants from other parts of town or from nearby underwater sediments can be dislodged by rushing floodwater and dumped on cultivated areas as floodwater recedes. Given wider climate change effects inducing greater frequencies and magnitudes of severe weather in many parts of the world, the geomorphological position of cultivated areas should be among the primary considerations in criteria used in urban food production planning (as there should be) and in political contestation relative to land access or distribution.

Urban soil conditions, as discussed previously, tend to be highly variable and may not be very conducive to cultivation. Much of this is due to prior and recurring human impact, but intrinsic soil properties, especially when largely unaltered by human action, are also the result of other factors. These can include texture and structure governing the flow and storage of water available for plant growth. There can also be greater amounts of trace elements to contamination and even pollution levels because of the kind of material out of which the soils formed. The extent to which such trace elements can contaminate vegetables and other crops or browsing domesticated animals also depends on the activities of microbes, who may make such trace elements more easily transferable to plant roots (Hursthouse and Leitão 2016).

These are but a few examples of how biophysical processes can affect urban cultivation, but the obverse is simultaneously true. Urban food production modifies environments and ecosystems. Cultivated areas, depending on their location, layouts, cropping system, and extent of permeable surface, change urban environments by mitigating urban heat island effects, increasing water infiltration rates and reducing flooding, potentially trapping dust along with attached contaminants, and providing more habitats and dispersal corridors, thereby enhancing biodiversity. Among other beneficial impacts, the overall amounts of organic wastes can be reduced through composting, as can be soil erosion with more and longer-lasting vegetation cover, and the impacts of food-related transport resulting from more localised consumption (Cilliers, Bouwman, and Drewes 2009, 100; Pataki et al. 2011). Except for the last item, these biophysically beneficial impacts are achieved by means of urban green spaces in general, and few studies compare the effects of different kinds of city areas covered by vegetation, cultivated or otherwise. The few that have point to relatively unsubstantial differences, except that cultivated urban soils may have higher soil organic carbon levels (when next to trees) and lower bulk density (i.e. less compaction and therefore less surface runoff). A major finding is the greater preservation of soil functions in urban green spaces compared to conventional farm and pastureland, at least in a UK area (Edmondson et al. 2014; Malinowska and Szumacher 2008). Overall, when it comes to soil quality, there seems to be insufficient evidence that urban cultivation is superior to other forms of green space.

In wider ecological terms, it might surprise that urban cultivation does not guarantee high marks in biodiversity. Firstly, gardens do not have to be cultivated for food to provide many and more amenable habitats than other parts of cities and even monoculture farmland. Home gardens have been noted for raising cultivar and some other plant diversity within as well as outside cities, but not necessarily for local plant species in total (Di Pietro et al. 2018; Galluzzi, Eyzaguirre, and Negri 2010; Gaston et al. 2005). This is not to downplay the ecological importance of urban cultivation and green space generally. Thousands of species may find homes within even a single garden, at least in temperate and tropical regions, thanks to a much greater amount of resources available, such as plentiful water, and milder temperatures and temperature ranges. Yet, for reasons discussed in Chapter 3 (e.g. high environmental heterogeneity, habitat fragmentation), the story is not so rosy for many species. Ants fare better in grassy vacant lots than in urban community gardens in Akron and Cleveland (Yadav, Duckworth, and Grewal 2012). For pollinators, as attested by a Portland, Oregon, case study, there still need to be much more data gathered about pollinator travelling habits and linkages between different parts of a landscape, within and beyond cities, to gauge the effects of urban cultivation plots on pollinators (Langellotto et al. 2018).

Nevertheless, long-term net primary productivity (the amount of yearly vegetation biomass) has been found to increase thanks to urban community gardens in several Eastern US cities (Parece and Campbell 2017b). Higher net primary productivity (e.g. the amount of biomass relative to local potential) is a sign of healthier ecosystems. Also, soil microbial populations and native species in general seem to prosper more frequently thanks to urban cultivation, despite the typically large number of introduced species, including those brought for food production purposes. Among the salient contributions to urban biodiversity are larger garden size, more diversified vegetation structure (e.g. multiple vegetation heights, clumping of trees), longer garden age, contiguity with high amounts of surrounding green space, and proximity of key external resources such as ponds (Gaston and Gaston 2011). The latter could be decisive for organisms like amphibians, whose life prospects could be improved (barring acid rain effects) by policies promoting an introduction or return to milpa food production systems in Central American cities and traditional wet rice systems in Southeast Asian cities (see Holzer et al. 2017). This has political implications, as do any urban ecosystem transformations. In this specific case, it means empowering economically and politically Indigenous communities in the Americas and peasant communities in Southeast Asia, where the know-how for these kinds of production systems remains strong despite historical attempts by colonisers, and later by neo-colonial or home-grown despotic regimes to undermine or annihilate them.

There are some aspects of food production in cities that could make for amplifying environmental problems unless the techniques used exclude recourse to conventional means. Farming, especially the fossil fuel-based industrialised variety, is among the major sources of water and soil destruction as well as biodiversity decline and GHG emissions, so an increase in food production in cities must also be met by great care about net environmental effects.

Water can be absorbed more readily and retained longer (in plants and soils) in green spaces generally, so urban cultivation can aid in water-conserving and flood magnitude-reducing endeavours. However, the net effects of irrigation for food production and of water-consuming food processing depend on climate conditions, the types of plants grown, the cultivation techniques used, and the source of water, among other factors. The matter remains understudied, including effects of the social conditions of cultivators on water use. With respect to soil, the tendency has so far been found to be in the main beneficial, at least in what can be verified for some temperate regions within the UK (Edmondson et al. 2014; Langemeyer, Latkowska, and Gómez-Baggethun 2016, 128–9).

The relationship between urban food production and GHG emissions is usually understood as inversely related, in that expanding urban cultivation should result in contributing to lowering total emissions. But this is not always clear, and there is insufficient evidence to support this claim for urban green spaces in general, as is claiming that there is any more effectiveness in reducing the levels of air and water pollutants (Egorov et al. 2016; Mok et al. 2014, 31–2; Pataki et al. 2011; Sarzhanov et al. 2017; Vasenev et al. 2018). Urban soils play a major role in governing total carbon fluxes in cities, but whether as net absorbers or emitters of carbon depends on soil properties as well as how they are impacted by people and soil-dwelling organisms. As described in Chapter 4, one must be mindful of organic and inorganic forms of carbon that can be stored in soils. In arid regions, most of the carbon is embedded in the inorganic forms, such as calcium carbonates. Dissolving carbonates with irrigation water, for example, can lead to CO_2 being emitted. In more humid regions, soil carbon is mainly in organic substances, and so anything that accelerates the breakdown of organic matter (such as ploughing) can result in more CO_2 going into the atmosphere. The net effects of urban cultivation may depend not only on regional climate but also on the methods of cultivation. In the London borough of Sutton, UK, Kulak, Graves, and Chatterton (2013) found that total CO_2 emissions in an urban community farm depend on the crops selected and whether or not they substitute for those with high carbon footprints (from transport). The type of urban soil cultivation technique, the kinds of plants and/or livestock selected, and the kinds of fertiliser used are of great consequence in terms of whether or not urban food production will contribute positively to reducing GHG emissions, or at least mitigating (if not neutralising) the net emissions linked to urban expansion.

There is also a tendency to forget that GHG emissions amount to much more than just CO_2, even if that is so far the most abundantly emitted compound. Other GHGs are related to cultivation and even more potent as contributors to global warming than CO_2, namely methane and nitrous oxide (Tian et al. 2015). The first is produced often by livestock and paddy rice, while the second results from inducing denitrifying bacteria to produce more nitrous oxide by ploughing and applying nitrogen fertiliser or animal wastes (Mason et al. 2017). Urban environments rarely include large enough spaces for wet rice cultivation, and it seems this is rarely practised within cities beyond tropical latitudes (or at best as small projects in very contained spaces; see https://modernfarmer.com/2014/02/rice-paddies-new-york-city/). However, in more amenable areas, especially in the tropics, wet rice can and is grown. Because of the small areas involved, the overall effect in methane production is likely insignificant compared to other sources within cities. However, dry rice cultivation, prominent among many West African farming communities, could be introduced in cities, so long as it makes sense ecologically and climatologically. Ruminants, another main source of methane, are also rarely integrated into urban food production, and they are likely of insufficient numbers to make much of a dent in terms of atmospheric emissions. The impact of various forms of methane cultivation needs further exploration, and to our knowledge, there is little to no research on the topic.

Nitrogen emissions, on the other hand, have been studied. The study by Edmondson et al. (2014), cited earlier, gives reason to claim that, at least in some humid and cool climate regions like the British Isles, nitrogen can be effectively cycled locally by means of urban cultivation and with minimal atmospheric NO_2 emissions. These positive effects are contingent on abstaining from nitrogen fertilisers and ensuring the use of animal wastes is done carefully and strictly monitored. How this would be politically achieved is also a crucial issue; an egalitarianism-promoting, assembly-based approach would be essential, we reckon. Otherwise, excluding people will result, among other ills, in reinforcing food-access disparities and continued distanciation from the food supply, aside from undermining any local food sovereignty prospects. Such efforts may anyway not offset nitrogen from vehicle emissions and other industrial sources, but they will be important in an urban future that dispenses with these sources of nitrogen emissions.

Looking at larger-scale capitalist processes may dampen the ecological outlook for urban cultivation, however. Privileging green spaces—including cultivated area—over built-up land, could lead to relocating activities with higher net carbon emissions to other places, especially suburbs and cities with regulatory frameworks that readily give in to land speculators (Glaeser and Kahn 2010). This means that vying for GHG emission reductions through urban food production may be an illusory undertaking as a result of displacement effects. This is a similar problem encountered with the introduction of materials to replace or cover contaminated soils. Without a clear grasp of off-site outcomes and inter-relations among places and without preventative internationally coordinated political struggles, what may benefit one place ecosocially comes at the expense of people and organisms elsewhere. The struggle for ecological sustainability must be as internationalist as the struggle for an egalitarian society.

Contaminants and urban cultivation

One of the most pressing concerns that cross (perhaps even unifies) many of the variegated circumstances discussed earlier is the incidence of highly concentrated toxic or potentially harmful substances. As noted in the previous chapter, industrialised and industrialising cities tend to be plagued by readily degradable as well as recalcitrant pollutants. For urban cultivation, contamination processes mean that working urban soils or eating city-grown produce can increase people's exposure to contaminants, especially in brownfields (areas that are or may be contaminated or polluted). The issue therefore merits special attention, given the potential health hazards involved.

For the most part, urban food productivity may be unimpeded by contaminants, unless they interfere with nutrient uptake, such as lead tying up calcium or phosphate, or by diminishing microbial populations that often sustain above-ground plant life. In part (see also Chapter 4), vegetables may continue to be prolific regardless of contaminant levels because the contaminant may also be a nutrient and also because the cultivar may be able to neutralise and thereby avoid or hyper-accumulate the substances without signs of health impairment (Ahmad 2016; Viehweger 2014; Yadav 2010). This is what makes contamination a particularly insidious problem, even if there are documented cases where toxicity levels are sufficiently high so as to reduce vegetable growth. The levels of trace elements, for example, may not impact plant growth much at all, but inhaling or ingesting particles laden with toxins will present a human health issue. Forcible closures of urban gardens, as happened in Montréal over the past decade (Platt 2014), may therefore have nothing to do with viability in terms of plant productivity in the strict sense. At the same time, as discussed below, one must also be wary of businesses and governments using contamination levels to justify land evictions, as happened in Sacramento (Cutts et al. 2017). Such pretexts can be easily exposed if the biophysical aspects and technical precautions are fully understood. This is one way in which the study and knowledge of biophysical processes is of direct political significance and, relative to the objectives we promote here, necessary in struggles for ecosocialism (see also Engel-Di Mauro 2020a).

The sources of contamination vary, and there are some complexities involved that need to be clarified to gain a better grasp of the problem. Soil particle re-suspension is an important, if under-appreciated, source of contamination in urban gardens (Clark, Brabander, and Erdil 2006; Wiseman, Zereini, and Püttmann 2015). Splash-derived re-deposition is another local factor (McBride et al. 2014). Airborne trace elements (such as heavy metals) are absorbed by leaves or can get lodged within other vegetable tissues (Schreck et al. 2012; Xiong et al. 2014). Brown, Chaney, and Hettiarachchi (2016) have called attention to local re-deposition of lead-laden soil particles, such that importing soil and creating raised beds may not suffice in contaminant exposure mitigation unless such interventions are continuously renewed (Clark, Hausladen, and Brabander 2008; Cooper et al. 2020). Long-term re-contamination monitoring policies, which, to our knowledge, exist nowhere yet, are essential in ensuring that re-contamination is avoided (for an overview of contamination from airborne sources, see Engel-Di Mauro 2020b).

The degree of contaminant entry varies according to element and vegetable type, and contamination by leaf absorption has been understudied relative to roots. Lead uptake relative to distance from soil surface also depends on the type of vegetable. Carrots absorb lead directly, but radishes and lettuce trap non-washable lead-laden particles. In contrast, tomatoes—because they are farther away from the soil surface—have shown negligible lead amounts in available studies (see also Cai, McBride, and Li 2016). Previous investigations by one of us, in several cities in New York State and in Rome, Italy (Engel-Di Mauro 2018, 2019), also suggest that atmospheric deposition may still be a source of contamination even in fruiting bodies (e.g. tomatoes and string beans), including those lodged in city-grown vegetables through splash and local re-deposition of soil particles laden with contaminants (Brevik and Burgess 2016, 73; Brown, Chaney, and Hettiarachchi 2016, 28; Clark and Knudsen 2013)

Air and water may also be contaminated, but strictly speaking, the problem for urban cultivation revolves about plants and soils, namely, what is added to them through wastewater irrigation, rainout and other airborne particle deposition, sewerage overflow, or other means. There is debate about the feasibility of growing safe food in cities and their health risk to growers. Some studies indicate that food and working conditions can be made safe, with qualifications and exceptions that are specific to site and crop. The matter is usually presented as a problem that can be overcome technically, using imported soil, avoiding toxin-accumulating crops, and other such prophylactic measures (Clark, Hausladen, and Brabander 2008; Gilmore 2001; Hendershot and Turmel 2007; Knapp et al. 2016; Säumel et al. 2012; Nabulo, Young, and Black 2010). Others suggest that the problem is much more difficult to avoid, especially in capital-poor countries, that is, most of the world, where wastewater use often introduces contaminants and pathogens. Furthermore, in cities, toxins are widespread, difficult to track (due to many diffuse and point sources), and constantly redistributed (Agbenin, Welp, and Danko 2010; Clark, Hausladen, and Brabander 2008; Kabata-Pendias 2011; Loynachan 2016). Some also question whether the contaminant exposure hazard is no different (if not lower) than other public health hazards common to cities, like air pollution from vehicular traffic and industrialised settings (Brown, Chaney, and Hettiarachchi 2016; Hursthouse and Leitão 2016, 142).

There are yet more factors to be considered that further complicate the story. Gardeners' inputs can introduce contaminants or alter soil properties in ways that favour vegetable contaminant absorption, depending on plant physiology (Bolan and Duraisamy 2003; De Miguel et al. 1998; Nabulo, Young, and Black 2010; Wortman and Lovell 2013). Different crops have diverse tolerance levels and bioaccumulation rates. Some crops, for instance in the *Brassica* genus (broccoli, cauliflower, turnip, and the like), are hyper-accumulators for some elements (Bourennane et al. 2010; Meuser 2010, 67). With so many variables involved, total soil trace element content (e.g. lead, nickel) does not necessarily correspond to higher vegetable trace element levels (Allen and Janssen 2006; Warming et al. 2015). This is crucial to grasp because soils with high levels of contaminants are typically deemed inappropriate for cultivation when the opposite may be true if precautions are put in place (see below for details). In urban gardens, soil-borne vegetable trace element contamination may also be infrequent because of typically high amounts of soil organic matter (Hursthouse and Leitão 2016; Kingery, Simpson, and Hayes 2001) or neutralising substances accidentally introduced from sources like construction debris (Howard and Olszewska 2011).

Transversal social causes of and biophysical constraints to urban cultivation

Notwithstanding the hype and exaggerated attention surrounding urban gardening in North America and North and Western Europe, most cities worldwide have been characterised by some form of food production for many decades or centuries, often since their very founding. However, it is also true that there has been a spike in urban food growing in the global North (though not necessarily its food productivity) over the past three decades. This spike is in part a result of pauperisation stemming from increased socioeconomic inequality, especially in countries with highest capital accumulations. Government austerity policies, supported by organs such as the International Monetary Fund, have made it possible for financial capitalism to recover from the 2008 Great Recession. Along with the hangover of depredations produced by deindustrialisation (Smith 1984), a new wave of pauperism has enhanced uneven development and its social damages—between and within cities.

In many cities in poor countries, urban food production is more like a form of smallholder agriculture with which households cope with worsening economic conditions. Often, farming is carried out in economically marginal and likely contaminated spaces, improvising with all sorts of techniques and cobbling of resources to compensate for poor or lacking infrastructure, as in Chongqing (see later). Contamination problems often involve recent and liberally dumped or emitted toxic substances, because of little regulation enforcement, and the spread of pathogens, often due to inadequate or missing sewage treatment plants. In humid climates with relatively even year-round precipitation, contaminants will largely be rained out and enter more readily into soils and crops. In cases of a preponderance of exposed or vegetated soils, contaminants may be more easily incorporated within soils and crops, or washed into rivers, lakes, or coastal waters because of flooding aided by compacted soils. Arid and semi-arid regions, and cities with little original vegetation cover left, will likely feature more constant redistributions of contaminants by means of particle (dust) movement.

Urban cultivation in the global South largely involves low-income households, in which the work is done mainly by women and, for the most part, manually. The areas farmed or cultivated are typically small, and the main motivation is to meet immediate needs or attenuate the effects of deprivation (Hovorka, de Zeeuw, and Njenga 2009). [Such social aspects are evident as well in many cities of the capital-accumulating countries (the global North), where community gardeners tend to be older females (see, for example, Diekmann, Gray, and Baker 2016).] When cultivation is more integrated with everyday livelihood activities, urban food producers will be exposed to contaminants daily and for long periods. Wastewater and pesticides, for instance, are commonly used in places like Tamale (Ghana; see next chapter), where piped water distribution is highly unequal, and economic survival has to be prioritised, in spite of concerns over hygiene (Bellwood-Howard and Bogweh Nchanji 2017, 86–7). The long-term health hazards likely resemble those of farmworkers generally and are probably greater than in other contexts of urban food production. At the same time, there are occasions where, as in Dar es Salaam (also discussed in the next chapter), urban cultivation not only may improve livelihoods but could also serve as a way to reduce flooding hazard exposure (Howorth, Convery, and O'Keefe 2001), unlike in global North places such as Naples or New York City, which are too paved-over for urban cultivation to make much of a dampening effect.

In cities where almost everyone is wage-dependent, urban food production is mainly confined to part-time operations that may or may not be community-based and that, for the most part, do not go beyond cultivation in terms of the amounts of food produced. In urban areas defined by long histories of manufacturing, environmental conditions may be hazardous for communities operating on brownfields, especially with contamination from long-term accumulations of organic and inorganic compounds or exposure to emissions from high vehicular traffic. In almost all cities, commercially minded operations, whether small or large, tend to be rare, while livestock-raising, if at all permitted, may or may not be combined with cropping. In some cities, there may also be many households linked to surrounding rural areas and thereby with access to land or pastures (Foeken 2006, 6–13; Mok et al. 2014; Orsini et al. 2013).

Some biophysical challenges exacerbate existing social inequalities in ways that are not altogether different from farming in the countryside. Rains may not correspond with times of greatest crop needs, or may be too little, causing drought; or excessive, causing floods. Soils in the latter case may become waterlogged and destroy crops that cannot withstand such conditions. There may be declining or poor soil fertility problems, pests, and diseases that are difficult to avoid or combat because of insufficient resources available. Some animal species may also consume or damage crops. In many situations, urban crop growers, like smallholders in the countryside, use agrochemicals sparingly because of having low incomes, but in towns the effects of agrochemicals and fertilisers are often magnified because urban settings tend to cover relatively smaller areas and to be peppered by many cavities and enclosed spaces that trap and concentrate substances (Foeken 2006, 10–11).

There exist other peculiarities mostly related to city life. Many of these are described in Chapter 4, and only some are highlighted here to give an idea of the obstacles facing urban food producers. Soils are often compacted, deficient in organic matter (hence low in major nutrients), too shallow relative to rooting depth, and erratic in terms of soil water conduction and retention, due to highly variable soil texture and structure (Beniston and Lal 2012). Raising livestock, a practice rare in the cities of capital-rich (ecologically indebted) countries, may increase the potential for street accidents, if transportation infrastructure and the main means of transport remain unchanged. There can also be problems of grazing in contaminated areas and of heightening soil erosion if soils are left without much vegetation cover (this can happen without any grazing, too). Tenuous land access also reduces investment into production, including agrochemical purchases and planting trees. In these kinds of circumstances, raising income levels and land tenure security through urban food production could have less than beneficial effects if growers elect to spend more money on agrochemicals and have no incentives or proper environments for trees. For places like Nakuru (Kenya), these challenges occur in a context of often violent policies undermining most people's livelihoods in the countryside. These policies are behind increasing migration to cities, urban area enlargement, and an expansion of urban food production in any area where land can be had, especially in former farmland at city edges (De Bon, Parrot, and Moustier 2010; Foeken 2006; Prain and Lee-Smith 2010). These are some of the major characteristics shared by city food production that also overlap with food production generally. Each situation in the end must be addressed in its full complexity and local specificity. The purpose of the generalisation here is only to summarise what is so far known and that can be helpful towards generating initial expectations of the kinds of challenges that may be prevalent as a result of biophysical conditions, as well as by city size and location, relative to its region and country.

Objectionable notions about society in the biophysical sciences

If only technical recommendations were fraught with problems, it would be feasible to arrive eventually at coherent overarching solutions by facilitating greater discussions among experts. The difficulty is much deeper, however. Experts in the biophysical sciences tend to make many unwarranted assumptions about society. They also often seem ingenuous or superficial about social causes and relations of power. At times, they show unwitting partisanship towards preserving or reinforcing existing social arrangements, especially liberal democracy (e.g. private ownership) and white colonial privilege. On rare occasions, allegiance to capitalist relations and even white supremacy are unabashedly flouted (see Clark 2016; Correia 2013). We concentrate here on urban social ecology, where, at times, arrogance towards and prejudice against marginalised peoples rear their ugly head with scientific veneer.

For instance, some ecologists seem to think that "wastelands which have been disused over long periods of time can provide valuable wilderness areas for recreation" and that such "wastelands" tend to coincide with "low-income residential areas" in "old industrial cities" (Pauleit and Breuste 2011, 30). In other instances, policy makers, informed by conventional notions of farming, deem it proper to draw up classification systems that deny the possibility of food production over thin, poorly drained soils on slopes exceeding 15 per cent (Sawio 1998, 22). This kind of perspective, based on no dialogue with local inhabitants and no data on actual uses, cultivation practices, and local knowledge, could not contrast more with the lived experiences of urban majorities, who tend to be low-income, with at least a third of global urban dwellers living in utterly abject conditions, often called slums (Davis 2006; Garland, Massoumi, and Ruble 2007). As discussed in the next chapter, in cities like Chongqing, cultivators can produce food even in the most unlikely places, on very poor soils or even by making soils. In many places, urban dwellers also know how to build terraces, compost organic wastes, and other such ameliorative techniques. What some ecologists and policy makers understand as "wastelands" or unsuitable soils are, actually, used or potentially usable spaces, sometimes appropriated by local inhabitants for self-provisioning. Immediate objectives range widely but, overall, they are responses to government and business actions, including land speculation and evictions (see Mudu and Marini 2018).

Other experts surmise that "biophysical issues . . . do not get enough emphasis in settled areas in developing countries, because socio-economic issues . . . are bigger concerns" (Cilliers, Bouwman, and Drewes 2009, 91). Problems like poverty and illiteracy in "developing countries" make "environmental awareness raising difficult to address" (ibid., 93). This is an oft-repeated refrain from technocrats that finds no support in reality (Guha and Martínez-Alier 1997; Moseley 2001) and is even contradicted, a few pages onwards, in the same authors' understanding that "environmental concerns are clearly linked to issues of poverty and social justice" (ibid., 101). Such linkages require little to no research for the satisfied scientists who use a single South African example to represent all developing countries. Yet it really would take but little effort to discover the myriad environmental justice movements present across South Africa. One might be forgiven for supposing that ecologists' deep illiteracy of environmentalism makes basic social justice awareness-raising, and possibly logical reasoning, difficult to address.

Potchefstroom, South Africa, presents an interesting case as an introduction of urban cultivation to disadvantaged communities. It is a municipality with a population of 128,000, located 120 km southwest of Johannesburg. A university city, Potchefstroom's claim to fame is being home to Fresh Pick, a community farm offering organic produce. With the best intentions, in the framework of the UN's Local Agenda 21 and Millennium Development Goals, an urban gardening project has been developed to help improve living conditions for poor Botswana, Xhosa, and South-Sotho peoples. However, the objectives of simultaneously improving living standards and ecosystem functioning soon clash with local understandings and practices. One such is the importance given to bare patches of land in front of houses to express cleanliness and another is the priority given to sharing resources and the product of labour with the rest of the community (the principle of *Ubuntu*). Rather than adapting urban cultivation projects accordingly, officials and scientists involved view these beliefs as impediments and unabashedly vie to impose capitalist practices (enticing vegetable sales, instead of communal redistribution) by means of "well-structured environmental education programmes" targeting, especially, women (Cilliers, Bouwman, and Drewes 2009, 104–5). The real aims of Millennium Development Goals, at least in this case, are therein gloriously revealed: conformity of all to the cardinal bourgeois principle of counting only that which can sell at profit (exchange-value). In some respects, this is a predictable outcome of a worldview in which ecology is to be subordinated to "planners and managers," whose business it is "to improve the conditions of the people in their care" (ibid. 109) with the sort of careless interventionism that refuses to look into causes of social inequalities and denies the possibility of community self-management. Such expert help, if heeded and applied as policy, will only exacerbate existing relations of domination. Therefore, not only are technical solutions being institutionally promoted incoherently relative to how to handle urban pollution affecting food production, but the very ideas about society among scientists must be dug up, questioned, and overcome. The building of biophysical knowledge is always as well a multi-faceted social struggle.

Underestimation and confusions about contamination

A situation where it is tough to parse through technical recommendations and technocratic ideologies is made even more difficult by those who downplay, if not omit, the topic of contamination, narrowing the focus to social aspects. Those who show awareness of the problem either underestimate it or seem to think it is resolvable simply by technical intervention (e.g. Hess 2009, 141; Knapp et al. 2016; Parece and Campbell 2017a, 38–9; Reynolds and Cohen 2016, 3; Schwarz et al. 2016; Silva and Pfeiffer 2016, 171). This is particularly perplexing when such views come from scholars who are otherwise wary of technocratic approaches. There is often as well a worrisome failure to comprehend even the basics among both scholars and activists, such as what needs to be sampled and tested (it is not just soils) and how trace elements like lead actually get into vegetables (e.g. Ladner 2011, 233–4; Tracey 2011, 140–1). Others limit themselves at most to describing the potential health risks, and then only a fraction of them. Typically, this is done without any analysis of preventative measures or of how pollution issues may affect the feasibility of urban food production (Knapp et al. 2016; Mok et al. 2014, 27–8; Orsini et al. 2013, 701). In studies that are not within the environmental sciences, where issues of oppression are largely omitted (or treated as if detached from politics and environmental impact), there is no prominence given to pollution problems (e.g. Gorgolewski, Kommisar, and Nasr 2011; Imbert 2015; Ladner 2011; Viljoen and Bohn 2014).

The salience and magnitude of the challenge are fortunately not lost on those in the biophysical sciences (e.g. Alloway 2013; Wortman and Lovell 2013) as well as some in the legal profession and social sciences (e.g. Gilmore 2001`; Platt 2014), who direct their efforts to analysing the problem and offer practicable technical solutions, including preventative actions. The trouble is that such solutions are divorced from the wider social context and ingenuous about relations of power (at any scale). For the most part, they lack any breakdown of who is most affected, how and why, and fixate on superficial or proximate causes (e.g. industrial emissions, urban growth, or improper soil management). What is even more sobering is that those directly involved in food production often do not know about potential risks (and their ultimate causes). Urban community gardens may be more effective than less inclusive kinds of gardening spaces in working against widespread alienation from the rest of nature (e.g. Bendt, Barthel, and Colding 2013), but they are hardly sufficient in developing ecological understandings of contamination processes. In studies conducted in Atlanta, Baltimore, and Chicago, it was found that cultivators are aware of the issue but tend not to know about sources and processes, exposure-prevention techniques, or how to interpret lab test results (Balotin et al. 2020; Hunter et al. 2020; Kim et al. 2014; Witzling, Wander, and Phillips 2011; see also Harms et al. 2013). There appears to be a yawning chasm between general assessments and practices of urban cultivation and the gravity of urban environmental challenges.

Similarly, and in tune with narrow scientific specialisation, a dialogue is missing among approaches on the fate and containment of contaminants and the environmental function of urban food production. Absorptive capacity benefits are disconnected from potential harms by exposure. On the one hand, cultivated spaces, like green spaces in general (Acosta et al. 2014), are hailed as neutralisers of contaminants by means of taking in and locking up toxins and breaking down their organic versions (Langemeyer, Latkowska, and Gómez-Baggethun 2016, 127–8; Pouyat et al. 2010, 124–6). On the other hand, concerns abound about contaminated soils as if they are unrelated from that praised neutralisation process (e.g. Ladner 2011, 233). We have then an unresolved and perhaps still unrecognised tension between using existing or creating new green spaces for long-term contaminant storage or for cultivation. However, it is not impossible technically to achieve both contaminant storage and food production on the same sites. However, for this to happen in ways that are safe for cultivators and the public at large, there must be a minimum guarantee of constant environmental monitoring (e.g. testing both produce and soils at least yearly), promptly available and pro-active (in person) technical advice, a mass educational campaign that is carried out on an egalitarian basis (i.e. having the same level of economic resources) among all directly concerned (for technical experts about social justice, for cultivators and the public at large about contamination processes), and a technical skilling process for cultivators so that they can grow food while neutralising any existing contaminants. For this to work, institutional policies must consist of excluding business interests altogether and of wealth redistribution, enabling community gardeners' involvement to be on equal terms with that of institutional technical experts. Such is part of the urban cultivation we envision, but it cannot be effectively developed and implemented in a stratified society like ours without simultaneously situating the cultivation process within an overarching struggle to arrive at an ecologically sustainable and egalitarian society via ecosocialism.

Furthermore, such biophysical assessments and monitoring would need to extend for decades to possibly hundreds of years (depending on initial contaminant levels and type of contaminants) until trace elements are sufficiently diluted to less than hazardous concentrations and/or until organic contaminants are degraded to harmless forms or levels. This lengthy and likely inter-generational process may see light at the end of the tunnel provided no further contamination happens in the meantime. Because there is still an overwhelming lack of institutional efficacy, resolve, or often even attention to contamination problems, prospects look rather grim, at the moment. They will continue to look grim until not just cities, but an entire social system is overhauled so that things are first and foremost produced to fulfil everyone's needs, not capital accumulation, and toxic emissions (if any) become practices of last resort, when all known alternatives have been exhausted (i.e. when the precautionary principle reigns).

All these matters related to contamination processes should be self-evidently political. That they are seldom so considered testifies to the overwhelming influence of capitalist ideologies, where politics are divorced from economics, or subordinated to primary economic (read capitalist) directives, and society and nature are separate or apart from each other (Smith 1984). Deciding over prioritising contaminant containment in, or cultivation of, contaminated soils or attempting to arrive at a judicious mixture of both is a political process in terms of how land is to be allocated and used. These are decision-making processes that unfold without much, if any, participation beyond that of experts and technocrats, who often answer to businesses' bottom line of profitability over ecosystem and public health. This accounts for what has already been under way without the sort of fanfare accorded to urban agriculture. What could be done to address the challenges properly is not yet on the political horizon. The process of re-educating technical experts, allocating funds towards monitoring, pro-active extension services, education, and cultivators' technical skilling, as part of developing lasting safeguards in the development of urban cultivation, necessitates awareness raising, political campaigning, and decisions made over funding priorities in cities. Still these would be but the first steps towards undoing the foundations of toxic emissions, that is, capitalist relations. Technical remedies can be of lasting beneficial consequence, ecologically and socially, if they are made part of a politics coherent with a shared vision of desired futures. In our case, we see ecosocialism as a desirable future and, like any political project, the extent to which such a vision is shared hinges on the relative success of social struggles. In this light, the usefulness of technical proposals and implementations like mitigation strategies is measured according to their consistency with a larger transformative political process leading to ecosocialist ends.

Ecosocial problems with conventional and alternative contaminant exposure mitigation

In the meantime, there are numerous ways to mitigate contaminant exposure that can be helpful towards overcoming the above challenges, if social justice issues and wider (and clearer) political objectives are taken to be as important as ecological sustainability. Some of these mitigation methods, happily, are already known to many cultivators (e.g. Tracey 2011, 139–41). These include removing and replacing entire soils or their most contaminated parts, importing soil periodically to replace soil surface layers, capping contaminated surfaces and introducing new soil on top, building and using raised beds, mulching surfaces to keep particles down, maintaining near-neutral pH levels, incorporating organic matter (e.g. compost), and thoroughly washing vegetables prior to consumption. Additionally, appropriate cultivars can be selected, avoiding tubers and hyper-accumulator species in areas with known high contaminant concentrations. Fruit and nut trees can make for an alternative source of food in such areas, since there tends to be very little contaminant transfer from roots to fruits (von Hoffen and Säumel 2014).

In some measure, the growing field of urban agroecology could be helpful to develop mitigating measures, boost ecological sustainability, and reduce social inequalities, at least within neighbourhoods where urban cultivation is introduced. To a large extent, these are potentials that are traceable to the sensibilities that emerged through tropical agroecology and, in part, organic farming movements, sensibilities that combine ecological with social inequality concerns in farming and wider food systems. It has been a response against the ecosocially devastating aspects of profit-centred, capital-intensive, mechanised, agrochemical and fossil-fuel dependent farming, the consequences of which have also contributed to massive rural-to-urban migration, huge food access inequalities, and enormous food waste discussed in the first two chapters (Gliessman 2013; Holt-Giménez 2017; Mies and Bennholdt-Thomsen 1999). Decades of research and learning from smallholder farming communities have led to basic recommendations for the development of agricultural techniques that rest on high biodiversity, preserving or increasing soil organic matter content, and enhancing nutrient cycling. These techniques have direct political and ecological ramifications as they help food producers avoid or minimise tillage, monocultures, agrochemicals (e.g. pesticides, synthetic fertilisers), and large-scale irrigation (especially in drylands). They raise smallholders' autonomy and greatly reduce, if not prevent, environmental degradation (e.g. soil erosion, genetic losses, GHG emissions, soil and water pollution, and habitat destruction) as well as perils to human health within farming communities and beyond. Agroecology, however, unlike most organic farming movements, calls into question not only industrial techniques, but also the capitalist relations that underpin the maldistribution of cultivable land and food, a primary basis of historical disenfranchisement and poverty for many agrarian communities (Altieri 2009; Gliessman 2015).

Nevertheless, like organic or sustainable agriculture, increasing institutional acceptance has coincided with technical and market-promoting perspectives overshadowing politically egalitarian ones (Méndez, Bacon, and Cohen 2016). Technocratic or conventional understandings of agroecology fail to consider (let alone explain) or provide any viable solutions to structural, long-term inequalities. Agroecologists who remain true to egalitarian objectives strive not only to establish methods that optimise (not necessarily maximise) food production in ecologically appropriate ways, but also to promote smallholder autonomy, using technical means to reduce or obviate industrial inputs and political strategies to enhance food system control. This implies securing access to land and food by attending to multiple-scale processes, thereby converging with political ecology (Altieri 1987; González de Molina 2016, 61). To achieve such simultaneously ecological and political aims, scientific research is combined with smallholders' knowledge systems by means of transdisciplinary participatory action approaches (Méndez, Bacon, and Cohen 2016).

Despite its complementarity with urban food justice, urban community gardening, etc., agroecology, as an explicit approach, remains largely peripheral to urban food production and alternative agri-food movements (Altieri and Nicholls 2018; Fernández et al. 2016; Silva and Pfeiffer 2016). Cuba is among the few exceptions where agroecological principles are formally integrated into urban food production projects, and we will explore this in the subsequent chapter with a focus on Havana (INIFAT 2010; Leitgeb et al. 2011). At the same time, to reiterate. many urban gardeners already practice agroecological principles without realising it (Hursthouse and Leitão 2016; Reynolds and Cohen 2016). The task should be anyway one of building on what gardeners already know and do, and this has, to some extent, occurred in Cuba.

Some may find the Cuban example indigestible for political reasons, but policies under capitalist conditions are even less democratic, as already shown for cities in the US and elsewhere. A showcase of urban agroecology, the city of Rosario, can illustrate the benefits and problems of urban agroecology in a liberal democratic setting. Located in Santa Fe Province, Argentina, 300 km northwest of Buenos Aires, Rosario's metropolitan area is home to 1.7 million. The city also hosts a partnership between the local government and the Resource Centre for Urban Agriculture and Forestry (RUAF), a foundation and international network dedicated to developing urban food production in developing countries. As briefly described earlier (Chapter 3), with the 2001 economic disaster in Argentina, women, for the most part, spearheaded an effort in Rosario to turn urban lots into food-producing areas, re-appropriating vacant municipal and private lots to grow food for their households (Ponce and Donoso 2009). The local government reacted with the 2002 Urban Agriculture Programme, in part rooted in earlier attempts during the late 1980s to establish agroecologically oriented gardens (Roitman and Bifarello 2010). As mainly a poverty alleviation measure, the programme is aimed at assisting the poor meet their dietary needs while offering economic opportunities (especially for women) and promoting agroecological growing methods. Impressively, agroecological vegetable and fruit production is currently integrated into official urban planning. The government provides district-level coordination (including establishing farmers' markets), technical and material assistance, and training. Hundreds of gardens have sprouted this way (Battiston et al. 2017; Guénette 2006; Lattuca 2017).

Astonishingly, given known pollution problems at the sites where gardens have been set up (Lattuca 2014, 87), there are no contamination assessments and no exposure-prevention measures being developed (apparently, not even discussed). In effect, the poor are encouraged to partake of a grossly underpaid clean-up operation without any safety precautions or equipment. Worse still, there are no proactive policies to improve living standards for the tens of thousands of Indigenous Toba (Qom) people in the city (Bigot 2007). Also, there is nothing on settler colonialism in either the municipal policy or the urban agroecology movement's projects (such as land restitution or Toba culture promotion and sensitisation). Importantly, the collectivism observed among at least some gardeners as well in neighbourhood organising for food production preceded the city's new policies (to some extent, it emerged as part of the abandoned factory occupations of the early 2000s). One could say that the municipality's Urban Agriculture Programme was therefore moulded to some degree from below as well. However, the policy explicitly privileges the commercialisation of agroecological production and the creation of entrepreneurs out of gardeners (Lilli 2017). The programme is also a way to retake control of land distribution and access, as the policy mainly applies to municipal lots. Absentee landlords get tax breaks for allowing gardeners to use such lots for two years, which means no long-term production planning (crucial for agroecological techniques to be workable) and no re-appropriation of privately owned land.

Taken together, conventional and alternative approaches and techniques (including agroecological) can help prevent or at least reduce exposure to local re-deposition, root-attachment of contaminant-bearing soil particles, and major processes of soil contaminant mobilisation leading to root absorption (e.g. Bellows 1999; EPA 2011; Hettariachchi et al. 2016; Hunter et al. 2020; Voigt and Leitão 2016). With respect to urban agroecology, techniques such as interplanting, appropriate crop sequencing, and maintaining constant soil cover (like permaculture, in some ways) have many ecological benefits, including raising productivity per unit of land, but the issues of native species suppression, contaminant containment, and net GHG emission reduction are not directly addressed yet, and no studies exist to test agroecological techniques in diverse urban situations to assess the overarching ecosocial effects. So far, urban agroecology studies are almost solely focused on social benefits, which mostly amount to stop-gap poverty reduction measures. The ecological positives are largely taken for granted, as if they magically followed once agroecological principles were put into practice (see, for example, Bowen Siegner, Acey, and Sowerwine 2020; RUAF 2017). In sum, conventional prevention measures and the still largely inchoate ideas and practices from urban agroecology are useful, but they fail to address wider ecological and social issues. In some ways, they may create more problems when actualised, if reliant on settler colonial relations and on (not unrelated) imported materials and tools, among other concerns.

The imported resources necessary to build and maintain urban cultivation areas are usually taken for granted. Their on-site ecological and social impacts, which tend to be beneficial, are not compared to the overall impacts on places out of which the materials originate. This is where LCA research becomes essential but insufficient since it is inattentive to relations of power in society and social relations generally. The net effects of urban cultivation may therefore not be as positive as claimed when the matter is viewed more broadly. They depend, among other factors, on what sort of inputs are used, their provenance, and how they are produced. Furthermore, soil removal only sends the contamination problem to other sites, for others to cope with in the end. Parisio (2018), for example, reports large amounts of contaminated material being transferred from New York City to parts of the Hudson Valley by means of illicit dumping operations. The phenomenon includes, to an unknown degree, the contaminated soil removed from sites turned into urban gardens. Similarly, large volumes of sediment, if not topsoil as well, are quarried and transferred to urban gardens as "clean soil." These destructive activities remain understudied and ignored by the many involved in urban cultivation.

Much has been made about bioremediation, which is one way of reducing contaminant exposure. For organic contaminants, specialised bacteria or fungi can be introduced to biodegrade substances into forms that do not harm us or most other species. In case of soil trace element abundance, one can plant vegetables other than accumulator species (e.g. lettuce, broccoli, and sunflower), root vegetables, and herbs (e.g. Hettariachchi et al. 2016; Kessler 2013). Fast-growing plant species and fungi that accumulate trace elements have been shown to be effective at reducing contaminant concentrations. Microbial (usually bacterial), worm, and fungal activities can be promoted successfully to degrade at least some organic contaminants, including those that are associated with petroleum production. Enzymes can also be injected into soils to aid in organic contaminant breakdown. The feasibility of these techniques depends on the amounts and type of contaminants in the soils, the size of the contaminated area, and on the possibility of matching introduced organisms or enzymes to local soil conditions, which may also change during the decontamination process.

In the former case, contaminant concentrations may be so high and the area affected so large that it would take a decade to more than a century for decontamination to be completed. What makes the issue complex is that soil characteristics may substantively change (e.g. pH may rise or more soil organic matter may be formed over time) so that, for instance, many trace elements may become unavailable for root absorption. At such junctures, recourse to chemical reagent applications may be necessary to make trace elements soluble so that they can flow into plant roots or fungal hyphae. In the latter case of matching organism or enzymes to soil, it may be difficult for organisms to thrive or for enzymes to be activated under given conditions, so that their populations or activity rates may be too low for effective organic contaminant breakdown or, in the case of fungi, trace element absorption. Chemical reagents can again be introduced to change soil properties like pH to help hasten the process of root or hyphal absorption of trace elements or organisms' or enzymes' biochemical breakdown of organic pollutants (Ghori et al. 2016).

The problem in all such cases is that some trace elements are more easily absorbed at high and some at low pH ranges and as soil conditions change so may the characteristics of some trace element contaminants. When a site is contaminated by multiple kinds of trace elements or when trace element properties change over time during the decontamination procedure, one is faced with deciding what trace element to attack first and with ensuring enough funding for long-term monitoring of soil and trace element characteristics so as to modify techniques as necessary. Some organic pollutants, like PCBs, are also too chemically inert to be attacked effectively by bacteria or enzymes, but some headway seems forthcoming using some fungal species. The by-products of organic contaminant breakdown, however, may also be toxic or toxic in high concentrations, like methane. Gases evolved from soils need to be carefully monitored and treated without displacing contamination to other places.

Potentials for displacement effects are one reason that bioremediation is, on the whole, not feasible, or, in some cases, not a solution at all. With respect to organic contaminants, there is the possibility of breaking down compounds to smaller, relatively harmless substances for most organisms. The breakdown process may take days to years, but there would not necessarily need to be any exporting of toxic substances to other places. Sometimes, as in the production of methane, the by-products of decontamination can be useful as local sources of energy (landfills, for instance, can also be used to such ends). The difficulty lies in the net GHG emissions involved and the amount of energy and materials needed to de-contaminate the site. Those resources largely come from somewhere else, so the multiplication of site de-contamination cases means that there needs to be great care about not exerting more demands for fossil fuel, quarrying, and mining extraction.

Trace elements are an altogether different matter that makes of bioremediation ultimately a non-starter. Organisms that concentrate high amounts of trace elements in their tissues must eventually be removed from site because, as they perish and are biodegraded, the trace elements will re-contaminate the soil. This implies that bioremediation of trace elements is another way of producing rubbish to be dumped somewhere else and, as environmental justice activists have long shown, on someone else with less political leverage. In part, the problem can be avoided using harvested plants as building or road materials or for other infrastructural works. But locking up concentrated hazardous materials this way only transforms a spatial into a temporal displacement effect. The eventual decay and destruction of such buildings, roads, and other such constructions will burden future generations with the contamination problem, much like buried radioactive waste does.

A more promising approach, under the rubric of "agro-mining" (see van der Ent et al. 2018), would be to extract the concentrated trace elements by incinerating the harvested plants so that the material can be re-used. In areas like those under cultivation in cities, which are usually tiny compared to farm fields in the countryside, the technique is likely not worth the effort. But the even greater impediment is deeply social in character. Extracting trace elements from highly polluted urban soils could be workable in an economy that was not based on capital accumulation (and for now, fossil fuels) and not founded on treating workers as expendable inputs. There would therefore first be the need of ensuring worker safety in the trace element extraction process, which implies strongly socialist unions at a minimum (explicitly anti-capitalist because otherwise unions become a transmission belt for the powers that be). Even if such a situation of worker power prevailed, anything extracted and recycled in this manner would be too little in volume to be deemed worthwhile compared to the amounts of trace elements extracted by means of existing mining operations. Even if the extraction through incineration were to become lucrative, the extent to which such recycling would be viable would be tied to often wild price fluctuations typical of capitalist systems.

These issues are difficult enough, but the overwhelming attention to soils as a vector of contaminant exposure has led to underestimating other sources. As already pointed out, soils are not the only conduit for contaminant transfer to vegetables. Nearby emissions (e.g. vehicular traffic, incinerators) and local gardening inputs can also contribute to contamination. This is complicated by the fact that contaminant influx occurs by means of various other kinds of atmospheric deposition. These include long-range wind transport (Hooda 2010, 4; McKenna-Neuman 2011) and local re-suspension of soil particles, which includes particle entrainment as well as splash (Alloway 2013, 25; Wortman and Lovell 2013). Soil particle re-suspension is known to occur within urban areas (Laidlaw and Filippelli

2008), and in a few cases, it has been recognised as an important source of trace element contamination in urban gardens (Clark, Brabander, and Erdil 2006; Engel-Di Mauro 2020b; Wiseman, Zereini, and Püttmann 2015). Splash-derived re-deposition is another local re-deposition factor (McBride et al. 2014). Airborne trace elements are absorbed by leaves or get lodged within other vegetable tissues (Schreck et al. 2012; Xiong et al. 2014). Brown, Chaney, and Hettiarachchi (2016) have called attention to local re-deposition of lead-laden soil particles; therefore, importing soil and creating raised beds may not suffice in trace element exposure mitigation (Clark, Hausladen, and Brabander 2008). Scraping and replacing the first few centimetres of soil, which can be done as a preventive measure, will only displace the contamination problem, as discussed earlier. The degree of contaminant entry in vegetables also varies according to trace element and vegetable type, and contamination by leaf absorption has been understudied relative to roots (Cai, McBride, and Li 2016; Paltseva et al. 2018). There is much that still needs to be researched and understood about what the main sources of contamination are, and which are more dangerous and in what situations. Existing recommendations, especially in the popular press and within the social sciences, give an impression of easily replicable recipe-like technical feasibility that is as misplaced as the occasional fear mongering about contaminants in the food produced in cities (for more details, see Engel-Di Mauro 2020a).

Political repercussions of biophysical processes in urban food production

One thing that can be said in favour of most experts in the biophysical sciences, irrespective of their usually tacit and often capitalism-friendly political positions, is that they are more attentive to the biophysical dimensions of life in the city. As already stated, promoters of urban food production typically leave out biophysical processes in their lofty disquisitions. This includes leftists who otherwise justifiably see in urban gardens the possibility for building post-capitalist egalitarian alternatives. Yet, without understanding the complexity of biophysical processes that affect the extent to which food production is feasible or the degree to which contaminants are dangerous, concerned movements or activists can underestimate health hazards and thereby, in the long run, undermine their own political projects. As contamination predominates and recurs as a main problem faced by urban food producers, we will emphasise here the social effects of enduring contamination. Ecosocially, conventional exposure-reduction effectiveness of preventive measures, like importing soil and building raised beds, cannot be regarded as recipes to be applied in any city. It is no simple task to establish the feasibility and political potentials of urban food production. Biophysical contexts, not only social ones, must be studied carefully to come up with ecologically sensible, health-preserving, as well as socially just alternatives. Contamination is determined by combined social and biophysical processes, so there are simultaneously political ramifications. There are political uses of instances of contamination and formidable obstacles to raising awareness and confronting them, as there are highly uneven knowledge bases and technological wherewithal among affected communities. Also, there are unappreciated ecological and social consequences of decontamination by displacement. There is the matter of typically unaddressed accountability intrinsic to pollution legacies. These are but some of the salient political implications of soil contamination (a thorough examination would require another book-length manuscript).

A lack of thorough understandings of biophysical processes can leave communities vulnerable to political attack. We have already alluded to one such example from Sacramento (Cutts et al. 2017). There, soil contamination was used in the 2004 local government's destruction of Mandella Community Garden. The local authorities' justification for the draconian measure was the presence of high amounts of lead. What was removed was not only all that the gardeners had built over several decades, but also between 30 and 120 cm of the soil they had cultivated (ibid., 14). The community garden was replaced with another one and, on virtually the same site, accompanied by an adjacent housing unit. The complex was renamed the Fremont Mews, endowed with 52 garden plots that people can access for a nominal fee. What is of special interest here is that local government and construction businesses appealed to soil lead contamination as a major justification to relocate the original community garden and, after much resistance, to destroy it altogether. They managed to succeed, as the authors put it, in "reframing the garden as dangerous dirt in need of purification" (ibid. 2017, 14). This brings to light the problem of activists not being in a position to counter institutional discourse about contamination with technical studies of their own and on their own terms. For instance, if activists had been privy to studies on soil lead and contamination prevention measures, they could have stated from the outset that lead contamination can be contained, and they would have been able to formulate their own contaminant management plans. This could have included raising pH levels and maintaining a permanent vegetation and mulch cover to restrain soil particle entrainment into the air (Menefee and Hettiarachchi 2018).

Few studies exist regarding urban cultivators' understandings of contamination issues. So far, as discussed earlier, they point to great difficulties in terms of even appreciating the existence of potential hazards, let alone knowledge of such processes or what to do about them, at least among Baltimore community gardeners (Kim et al. 2014). Elsewhere, as in Atlanta, gardeners are well-aware of contamination potentials, but are frustrated by barriers like inadequate funding and a dearth of training opportunities (Hunter et al. 2020). This is in spite of readily available resources from extension agencies in the US, where much effort has been expended in developing plain-language explanations and recommendations for effective exposure prevention (for a Baltimore-specific example, see also Johns Hopkins Center for a Livable Future 2014). Barriers to urban cultivators' understanding of soil contamination processes point to highly uneven knowledge bases and technological wherewithal (e.g. field equipment, laboratories) in most affected communities. There is an unacknowledged politics of knowledge among even the most well-meaning officials and technical experts about access to information, studies, education, and the infrastructure needed to conduct sampling and analysis. These essential ingredients to determining the presence and degree of contamination and implementing appropriate measures are typically denied socially in gendered and racialised ways. This is particularly consequential because, worldwide, most urban food production is carried out by women and people of colour (Mougeot 2015; Nyantakyi-Frimpong, Arku, and Baah Inkoom 2016; Orsini et al. 2013; Reynolds and Cohen 2016).

Relative to ecological and social displacement, site decontamination is increasingly involving the translocation of ecological ills to other communities. Where contaminated sediment is dumped, and which places will experience soil and sediment losses seem not to have even entered the minds of concerned researchers and activists. In instances of contaminated material transfer, affected communities are not being informed about such hazards, least of all approached regarding their consent. The matter involves private property and/or technocratic state organs, which exclude public involvement unless legal actions are taken (i.e. when damage is already done). Such geographically wider ecological and social repercussions of the spread of urban cultivation, especially in the most capitalised cities of North America and Western Europe, continue to be unaccounted for or ignored.

There are even more profound issues further confounding prospects for accountability, at least as widely understood in capitalist systems. Trace elements (e.g. lead) and other persistent contaminants often long surpass the longevity of those most responsible for producing the problem. The issue of responsibility for environmental harms is the most pressing yet hardly resolved political issue, or, arguably, resolvable under current political systems. Leaded petrol and paint, for example, involved multiple businesses, from production to distribution to consumption. Lead remains among the most widespread pollutants in all major cities and beyond (Li et al. 2017; Mitchell et al. 2014), so most urban cultivation projects must become knowledgeable of lead contamination potential, find ways of testing soils and produce for lead, and learn ways to circumvent contaminant exposure. To date no businesses have been called to task to pay for all the damage caused, still causing, and still to be caused. This is besides the fact that, from the beginning, alternatives to leaded petrol and paint and their dangerous health consequences were well known (Kitman 2000). If responsibility were seriously considered and legally pursued (which is not the case under current political conditions), it would be impossible for any such businesses to be able to afford clean-up costs, present and future healthcare bills, and any other conventionally understood costs, never mind social harms that cannot be rectified monetarily. In the case of neurological damage by lead, one cannot simply buy a new brain package including all the social life experiences that produced it. Another aspect is the diffuse aspect of lead pollution, which goes beyond ultimately responsible but as yet unaccountable businesses. This aspect involves construction and associated soil erosion, leading to lead-bearing particles being redistributed by water and wind within and beyond cities. Finding responsibility for these kinds of physical and social processes would be most arduous and likely counterproductive.

It would entail blaming almost every business in existence. Complicating the matter further is the fact that the people most responsible may be deceased.

Even if one were able to find an original and still living culprit and wage a successful legal battle, one would still fail to address the lasting damage done and prevent similarly or even more destructive processes in future. As Joel Kovel (2002) pointed out in the case of the massive 1984 Bhopal pesticide factory leakage, which killed or maimed tens of thousands, apprehending Union Carbide (now Dow Chemical) officials will not put much of a dent on the underlying generative process that brings about such devastation. This process is the propensity for capitalist relations to undermine the conditions of existence for most of us and for many species in their entirety (O'Connor 1988). Not even localised long-lasting pollution effects are resolvable by identifying the culpable, who, in the case of Union Carbide (and is so many other such cases), have been able to evade responsibility altogether. Since the causes are inexorably systemic in nature, a systemic change is necessary, one that moves beyond bourgeois notions of individual culpability to one that recognises the need for substantive social control over the ways the economy is set up and run relative to processes like land use. Instead of socialising costs or re-directing them to the next owner, the problem of diffuse or unaccountable causality should be seized as an opportunity to argue for the devolution of urban planning and management to all urban inhabitants. As Lefebvre (1974), for one, argued, this could lead to reclaiming entire cities as commons rather than continuing with the current piecemeal approach to land struggles subtending urban food production today.

References

Acosta, J.A., A. Faz, S. Martínez-Martínez, and J.M. Arocena. 2014. Grass-Induced Changes in Properties of Soils in Urban Green Areas with Emphasis on Mobility of Metals. *Journal of Soils and Sediments* 14: 819–28.

Agbenin, J.O., G. Welp, and M. Danko. 2010. Fractionation and Prediction of Copper, Lead, and Zinc Uptake by Two Leaf Vegetables from Their Geochemical Fractions in Urban Garden Fields in Northern Nigeria. *Communications in Soil Science and Plant Analysis* 41: 1028–41.

Ahmad, P., ed. 2016. *Plant Metal Interaction*. Amsterdam: Elsevier.

Allen, H.E., and C.R. Janssen. 2006. Incorporating Bioavailability into Criteria for Metals. In, *Soil and Water Pollution Monitoring, Protection, and Remediation*, edited by I. Twardowska, H.E. Allen, and M.A. Häggblom, 93–105. Dordrecht: Springer.

Alloway, B.J. 2013. Sources of Heavy Metals and Metalloids in Soils. In, *Heavy Metals in Soils. Trace Metals and Metalloids in Soils and Their Bioavailability*, edited by B.J. Alloway, 3rd ed., 11–50. London: Chapman & Hall.

Altieri, M.A. 1987. *Agroecology: The Scientific Basis of Alternative Agriculture*. Boulder, CO: Westview Press.

Altieri, M.A. 2009. Agroecology, Small Farms, and Food Sovereignty. *Monthly Review* 61(3): 102–13.

Altieri, M.A., and C. Nicholls. 2018. Urban Agroecology: Designing Biodiverse, Productive and Resilient City Farms. *Agro Sur* 46(2): 49–60.

Balotin, L., S. Distler, A. Williams, S.J.W. Peters, C.M. Hunter, C. Theal, G. Frank, T. Alvarado, R. Hernandez, A. Hines, and E. Saikawa. 2020. Atlanta Residents' Knowledge Regarding Heavy Metal Exposures and Remediation in Urban Agriculture. *International Journal of Environmental Research and Public Health* 17: 2069.

Barbeau, C.D., M.J. Wilton, M. Oelbermann, J.D. Karagatzides, and L.J.S. Tsuji. 2018. Local Food Production in a Subarctic Indigenous Community: The Use of Willow (*Salix* spp.) Windbreaks to Increase the Yield of Intercropped Potatoes (*Solanum tuberosum*) and Bush Beans (*Phaseolus vulgaris*). *International Journal of Agricultural Sustainability* 16(1): 29–39.

Battiston, A., G. Porzio, N. Budai, N. Martínez, Y. Pérez Casella, R. Terrile, M. Costa, A. Mariatti, and N. Paz. 2017. Green Belt Project: Promoting Agroecological Food Production in Peri-Urban Rosario. *Urban Agriculture Magazine* 33: 51–2.

Bellows, A.C. 1999. Urban Food, Health and the Environment: The Case of Upper Silesia, Poland. In, *For Hunger-Proof Cities: Sustainable Urban Systems*, edited by M. Koc, R. MacRae, L.J.A. Mougeot, and J. Welsh, 131–5. Ottawa: IDRC.

Bellwood-Howard, I., and E. Bogweh Nchanji. 2017. The Marketing of Vegetables in a Northern Ghanaian City: Implications and Trajectories. In, *Global Urban Agriculture: Convergence of Theory and Practice between North and South*, edited by A.M.G.A. WinklerPrins, 79–92. Boston, MA: CABI.

Bendt, P., S. Barthel, and J. Colding. 2013. Civic Greening and Environmental Learning in Public-Access Community Gardens in Berlin. *Landscape & Urban Planning* 109(1): 18–30.

Beniston, J., and R. Lal. 2012. Improving Soil Quality for Urban Agriculture in the North Central U.S. In, *Carbon Sequestration in Urban Ecosystems*, edited by R. Lal, and B. Augustin, 279–313. Dordrecht: Springer.

Bigot, M. 2007. *Los Aborígenes Qom en Rosario: Contacto Lingüístico-Cultural, Bilingüismo, Diglosia y Vitalidad Etnolingüística en Grupos Aborígenes "Qom" (Tobas) Asentados en Rosario*. Rosario: UNR Editora.

Bolan, N.S., and V.P. Duraisamy. 2003. Role of Inorganic and Organic Soil Amendments on Immobilisation and Phytoavailability of Heavy Metals: A Review Involving Specific Case Studies. *Australian Journal of Soil Research* 41(3): 533–55.

Botkin, D.B., and C.E. Beveridge. 1997. Cities as Environments. *Urban Ecosystems* 1: 3–19.

Bourennane, H., F. Douay, T. Sterckeman, E. Villanneau, H. Ciesielski, D. King, and D. Baize. 2010. Mapping of Anthropogenic Trace Elements Inputs in Agricultural Topsoil from Northern France Using Enrichment Factors. *Geoderma* 157: 165–74.

Bowen Siegner, A., C. Acey, and J. Sowerwine. 2020. Producing Urban Agroecology in the East Bay: From Soil Health to Community Empowerment. *Agroecology and Sustainable Food Systems* 44(5): 566–93.

Brevik, E.C., and L.C. Burgess. 2016. *Soils and Human Health*. Boca Raton, FL: CRC Press.

Brookfield, H. 2001. *Exploring Agrodiversity*. New York: Columbia University Press.

Brown S.L., R.L. Chaney, and G.M. Hettiarachchi. 2016. Lead in Urban Soils: A Real or Perceived Concern for Urban Agriculture? *Journal of Environmental Quality* 45: 26–36.

Cai, M., M.B. McBride, and K. Li. 2016. Bioaccessibility of Ba, Cu, Pb, and Zn in Urban Garden and Orchard Soils. *Environmental Pollution* 208(Part A): 145–52.

Childers, D.L., M.L. Cadenasso, J.M. Grove, V. Marshall, B. McGrath, and S.T.A. Pickett. 2015. An Ecology *for* Cities: A Transformational Nexus of Design and Ecology to Advance Climate Change Resilience and Urban Sustainability. *Sustainability* 7: 3774–91.

Cilliers, S., H. Bouwman, and E. Drewes. 2009. Comparative Ecological Research in Developing Countries. In, *Ecology of Cities and Towns: A Comparative Approach*, edited by M.J. McDonnell, A.K. Hahs, and J.H. Breuste, 90–111. Cambridge: Cambridge University Press.

Clark, H.F., D.J. Brabander, and R.M. Erdil. 2006. Sources, Sinks, and Exposure Pathways of Lead in Urban Garden Soil. *Journal of Environmental Quality* 35: 2066–74.

Clark, H.F., D.M. Hausladen, and D.J. Brabander. 2008. Urban Gardens: Lead Exposure, Recontamination Mechanisms, and Implications for Remediation Design. *Environmental Research* 107: 312–19.

Clark, J.J. 2016. *The Tragedy of Common Sense*. Regina: Changing Suns Press.

Clark, J.J., and A.C. Knudsen. 2013. Extent, Characterization, and Sources of Soil Lead Contamination in Small-Urban Residential Neighborhoods. *Journal of Environmental Quality* 42: 1498–506.

Clucas, B., and J.M. Marzluff. 2011. Coupled Relationships Between Humans and Other Organisms in Urban Areas. In, *Urban Ecology. Patterns, Processes, and Applications*, edited by J. Niemelä, 135–47. Oxford: Oxford University Press.

Cooper, A.M., D. Felix, F. Alcantara, I. Zaslavsky, A. Work, P.L. Watson, K. Pezzoli, Q. Yu, D. Zhu, A.J. Scavo, Y. Zarabi, and J.I. Schroeder. 2020. Monitoring and Mitigation of Toxic Heavy Metals and Arsenic Accumulation in Food Crops: A Case Study of an Urban Community Garden. *Plant Direct* 4: 1–12.

Correia, D. 2013. F**k Jared Diamond. *Capitalism Nature Socialism* 24(4): 1–6.

Cutts, B.B., J.K. London, S. Meiners, K. Schwarz, and M.L. Cadenasso. 2017. Moving Dirt: Soil, Lead, and the Dynamic Spatial Politics of Urban Gardening. *Local Environment* 22(8): 998–1018.

Davis, M. 2006. *Planet of Slums*. London: Verso.

De Bon, H., L. Parrot, and P. Moustier. 2010. Sustainable Urban Agriculture in Developing Countries. A Review. *Agronomy for Sustainable Development* 30(1): 21–32.

De Miguel, E., M. Jimenez de Grado, J.F. Llamas, A. Martin-Dorado, and L.F. Mazadiego. 1998. The Overlooked Contribution of Compost Application to the Trace Element Load in the Urban Soil of Madrid (Spain). *The Science of the Total Environment* 215: 113–22.

Diekmann, L., L. Gray, and G. Baker. 2016. *Food Gardening in Santa Clara County*. Santa Clara, CA: Food and Agribusiness Institute, Santa Clara University.

Di Pietro, F., L. Mehdi, M. Brun, and C. Tanguay C. 2018. Community Gardens and Their Potential for Urban Biodiversity. In, *The Urban Garden City. Cities and Nature*, edited by S. Glatron, and L. Granchamp, 131–51. Heidelberg: Springer Cham.

Edmondson, J.L., Z.G. Davies, K.J. Gaston, and J.R. Leake. 2014. Urban Cultivation in Allotments Maintains Soil Qualities Adversely Affected by Conventional Agriculture. *Journal of Applied Ecology* 51: 880–9.

Egorov, A.I., P. Mudu, M. Braubach, and M. Martuzzi, eds. 2016. *Urban Green Spaces and Health. A Review of the Evidence*. Copenhagen: WHO Regional Office for Europe.

Engel-Di Mauro, S. 2018. An Exploratory Study of Potential as and Pb Contamination by Atmospheric Deposition in Two Urban Vegetable Gardens in Rome, Italy. *Journal of Soils and Sediments* 18(2): 426–30.

Engel-Di Mauro, S. 2019. Urban Vegetable Garden Soils and Lay Public Education on Soil Heavy Metal Exposure Mitigation. In, *Green Technologies and Infrastructure to Enhance Urban Ecosystem Services. Proceedings of the Smart and Sustainable Cities Conference* 2018, edited by V. Vasenev, E. Dovletyarova, Z. Cheng, R. Valentini, and C. Calfapietra, 221–6. Cham: Springer.

Engel-Di Mauro, S. 2020a. The Troubling and Troublesome Worlds of Urban Soil Trace Element Contamination Baselines. *Environment and Planning E: Nature and Space* 3(1): 95–113.

Engel-Di Mauro, S. 2020b. Atmospheric Sources of Trace Element Contamination in Cultivated Urban Areas: A Review. *Journal of Environmental Quality*. https://doi.org/10.1002/jeq2.20078.

EPA. 2011. Brownfields and Urban Agriculture. Interim Guidelines for Safe Gardening Practices. www.epa.gov/sites/production/files/2015-09/documents/bf_urban_ag.pdf (Accessed 7 August 2020).

Fernández, M., V.E. Méndez, T. Mares, and R. Schattman. 2016. Agroecology. Food Sovereignty, and Urban Agriculture in the United States. In, *Agroecology. A Transdisciplinary, Participatory and Action-Oriented Approach*, edited by V.E. Méndez, C.M. Bacon, R. Cohen, and S.R. Gliessman, 161–75. Boca Raton, FL: CRC Press.

Foeken, D. 2006. *To Subsidise My Income: Urban Farming in an East-African Town*. Boston, MA: Brill Academic Publishers.

Galluzzi, G., P. Eyzaguirre, and V. Negri. 2010. Home Gardens: Neglected Hotspots of Agro-Biodiversity and Cultural Diversity. *Biodiversity Conservation* 19: 3635–54.

Garland, A.M., M. Massoumi, and B.A. Ruble, eds. 2007. *Global Urban Poverty: Setting the Agenda*. Washington, DC: Woodrow Wilson International Center for Scholars.

Gaston, K.J., and S. Gaston. 2011. Urban Gardens and Biodiversity. In, *The Routledge Handbook of Urban Ecology*, edited by I. Douglas, D. Goode, M. Houck, and R. Wang, 450–8. London: Routledge.

Gaston, K.J., P.H. Warren, K. Thompson, and R-M. Smith. 2005. Urban Domestic Gardens (IV): The Extent of the Resource and Its Associated Features. *Biodiversity Conservation* 14: 3327–49.

Ghori, Z., H. Iftikhar, M.F. Bhatti, N. um-Minullah, I. Sharma, A.G. Kazi, and P. Ahmad. 2016. Phytoextraction: The Use of Plants to Remove Heavy Metals from Soil. In, *Plant Metal Interaction*, edited by P. Ahmad, 385–409. Amsterdam: Elsevier.

Gilmore, E. 2001. A Critique of Soil Contamination and Remediation: The Dimensions of the Problem and the Implications for Sustainable Development. *Bulletin of Science Technology & Society* 21: 394–400.

Glaeser, E.L., and M.E. Kahn. 2010. The Greenness of Cities Carbon Dioxide Emissions and Urban Development. *Journal of Urban Economics* 67: 404–18.

Gliessman, S.R. 2013. Agroecology: Growing the Roots of Resistance. *Agroecology and Sustainable Food Systems* 37: 19–31.

Gliessman, S.R. 2015. *Agroecology. The Ecology of Sustainable Food Systems*, 3rd ed. Boca Raton, FL: CRC Press.

González de Molina, M. 2016. Political Agroecology. An Essential Tool to Promote Agrarian Sustainability. In, *Agroecology. A Transdisciplinary, Participatory and Action-Oriented Approach*, edited by V.E. Méndez, C.M. Bacon, R. Cohen, and S.R. Gliessman, 55–72. Boca Raton, FL: CRC Press.

Gorgolewski, M., J. Kommisar, and J. Nasr. 2011. *Carrot City. Creating Places for Urban Agriculture*. Singapore: The Monacelli Press.

Guénette, Louise. 2006. A City Hooked in Urban Farming. IDRC. www.idrc.ca/sites/default/files/sp/Documents%20EN/city-hooked-on-urban-farming.pdf (Accessed 23 July 2020).

Guha, R., and J. Martínez-Alier. 1997. *Varieties of Environmentalism. Essays North & South*. London: Earthscan.

Harms, A.M.R., D.R. Presley, G.M. Hettiarachchi, and S.J. Thien. 2013. Assessing the Educational Needs of Urban Gardeners and Farmers on the Subject of Soil Contamination. *Journal of Extension* 51(1). www.joe.org/joe/2013february/ a10.php (Accessed 1 August 2018).

Harvey, D. 1996. *Justice, Nature, and the Geography of Difference*. Oxford: Blackwell.

Hendershot, W., and P. Turmel. 2007. Is Food Grown in Urban Gardens Safe? *Integrated Environmental Assessment and Management* 3: 463–4.

Hess, D.J. 2009. *Localist Movements in a Global Economy. Sustainability, Justice, and Urban Development in the United States*. Cambridge: The MIT Press.

Hettariachchi, G.M., C.P. Attanayake, P.P. Defoe, and S.E. Martin. 2016. Mechanisms to Reduce Risk Potential. In, *Sowing Seeds in the City. Human Dimensions*, vol. 2, edited by E. Hodges Snyder, K. McIvor, and S. Brown, 155–70. Dordrecht: Springer.

Holt-Giménez, E. 2017. *A Foodie's Guide to Capitalism: Understanding the Political Economy of What We Eat*. New York: Monthly Review.

Holzer, K.A., R.P. Bayers, T.T. Nguyen, and S.P. Lawler. 2017. Habitat Value of Cities and Rice Paddies for Amphibians in Rapidly Urbanizing Vietnam. *Journal of Urban Ecology* 3(1). https://doi.org/10.1093/jue/juw007 (Accessed 7 July 2020).

Hooda, P., ed. 2010. *Trace Elements in Soils*. Oxford: Blackwell.

Hovorka, A., H. de Zeeuw, and M. Njenga. 2009. Gender in Urban Agriculture: An Introduction. In, *Women Feeding Cities: Mainstreaming Gender in Urban Agriculture and Food Security*, edited by A. Hovorka, H. de Zeeuw, and M. Njenga, 1–32. Warwickshire: Practical Action Publishing.

Howard, J.L., and D. Olszewska. 2011. Pedogenesis, Geochemical Forms of Heavy Metals, and Artifact Weathering in an Urban Soil Chronosequence, Detroit, Michigan. *Environmental Pollution* 159(3): 754–61.

Howorth, C., I. Convery, and P. O'Keefe. 2001. Gardening to Reduce Hazard: Urban Agriculture in Tanzania. *Land Degradation & Development* 1(3): 285–91.

Hunter, C.M., D.H.Z. Williamson, M. Pearson, E. Saikawa, M.O. Gribble, and M. Kegler. 2020. Safe Community Gardening Practices: Focus Groups with Garden Leaders in Atlanta, Georgia. *Local Environment* 25(1): 18–35.

Hursthouse, A.S., and T.E. Leitão. 2016. Environmental Pressures on and the Status of Urban Allotments. In, *Urban Allotment Gardens in Europe*, edited by S. Bell, R. Fox-Kämper, N. Keshavarz, M. Benson, S. Caputo, S. Noori, and A. Voigt, 142–64. London: Routledge and Earthscan.

Imbert, D., ed. 2015. *Food and the City, Histories of Culture and Cultivation*. Washington, DC: Dumbarton Oaks.

INIFAT (Instituto de Investigaciones Fundamentales en Agricultura Tropical). 2010. *Manual Técnico para Organopónicos, Huertos Intensivos y Organoponía Semiprotegida*. Septima Edición. Habana: INIFAT.

Johns Hopkins Center for a Livable Future. 2014. Soil Safety Resource Guide for Urban Food Growers. www.jhsph.edu/research/centers-and-institutes/johns-hopkins-center-for-a-livable-future/_pdf/projects/urban-soil-safety/CLF%20Soil%20Safety%20Guide.pdf (Accessed 30 April 2021).

Kabata-Pendias, A. 2011. *Trace Elements in Soils and Plants*, 4th ed. Boca Raton, FL: CRC Press.

Kessler, R. 2013. Urban Gardening: Managing the Risks of Contaminated Soil. *Environmental Health Perspectives* 121: A326–A333.

Kim, B.F., M.N. Poulsen, J.D. Margulies, K.L. Dix, A.M. Palmer, and K.E. Nachmann. 2014. Urban Community Gardeners' Knowledge and Perceptions of Soil Contaminant Risks. *PLoS One* 9(2): e87913.

Kingery, W.L., A.J. Simpson, and M.H.B. Hayes. 2001. Chemical Structures of Soil Organic Matter and Their Interactions with Heavy Metals. In, *Heavy Metals Release in Soils*, edited by H.M. Selim, and D.L. Sparks, 237–45. Boca Raton, FL: Lewis Publishers.

Kitman, J.L. 2000. The Secret History of Lead. *The Nation* 26: 11–44.

Knapp, L., E. Veen, H. Renting, H. Wiskerke, and J. Groot. 2016. Vulnerability Analysis of Urban Agriculture Projects: A Case Study of Community and Entrepreneurial Gardens in the Netherlands and Switzerland. *Urban Agriculture & Regional Food Systems* 1(1): 1–13.

Kovel, J. 2002. *The Enemy of Nature. The End of Capitalism or the End of the World?* London: Zed Books.

Kulak, M., A. Graves, and J. Chatterton. 2013. Reducing Greenhouse Gas Emissions with Urban Agriculture: A Life Cycle Assessment Perspective. *Landscape and Urban Planning* 111: 68–78.

Ladner, P. 2011. *The Urban Food Revolution. Changing the Way We Feed Cities.* Babriola Island: New Society Publishers.

Laidlaw, M.A.S., and G.M. Filippelli. 2008. Resuspension of Urban Soils as a Persistent Source of Lead Poisoning in Children: A Review and New Directions. *Applied Geochemistry* 23: 2021–39.

Langellotto, G.A., A. Melathopoulos, I. Messer, A. Anderson, N. McClintock, and L. Costner. 2018. Garden Pollinators and the Potential for Ecosystem Service Flow to Urban and Peri-Urban Agriculture. *Sustainability* 10: 2047.

Langemeyer, J., M.J. Latkowska, and E.N. Gómez-Baggethun. 2016. Ecosystem Services from Urban Gardens. In, *Urban Allotment Gardens in Europe*, edited by S. Bell, R. Fox-Kämper, N. Keshavarz, M. Benson, S. Caputo, S. Noori, and A. Voigt, 115–41. London: Routledge and Earthscan.

Lattuca, Antonio. 2014. Rosario. In, *Growing Greener Cities in Latin America and the Caribbean*, edited by Graeme Thomas, 80–9. Rome: FAO.

Lattuca, Antonio. 2017. Using Agroecological and Social Inclusion Principles in the Urban Agriculture Programme in Rosario, Argentina. *Urban Agriculture Magazine* 33: 51–2.

Lefebvre, Henri. 1974 [1991]. *The Production of Space.* Translated by D. Nicholson-Smith. Oxford: Blackwell.

Leitgeb, F., F.R. Funes-Monzote, S. Kummer, and C.R. Vogl. 2011. Contribution of Farmers' Experiments and Innovations to Cuba's Agricultural Innovation System. *Renewable Agriculture and Food Systems* 26(4): 354–67.

Levins, R., and R. Lewontin. 1985. *The Dialectical Biologist.* Cambridge: Harvard University Press.

Li, I., Z. Cheng, A. Paltseva, T. Morin, B. Smith, and R. Shaw. 2017. Lead in New York City Soils. In, *Megacities 2050: Environmental Consequences of Urbanization: Proceedings of the VI International Conference on Landscape Architecture to Support City Sustainable Development*, edited by V.I. Vasenev, E. Dovletyarova, Z. Cheng, and R. Valentini, 62–79. Berlin: Springer Geography.

Lilli, L. 2017. 'Sembrar lo Colectivo': La Participación en la Red de Huerteros y Huerteras en la Ciudad de Rosario 2005–2015. *REA* (Escuela de Antropología FHUMYAR-UNR) XXIII: 209–28.

Loynachan, T.E. 2016. Human Disease from Introduced and Resident Soilborne Pathogens. In, *Soils and Human Health*, edited by E.C. Brevik, and L.C. Burgess, 107–36. Boca Raton, FL: CRC Press.

Malinowska, E., and I. Szumacher. 2008. Rola OgodówDziałkowych w Krajobrazie Lewobrzeżnej Warszawy. [Role of Allotment Gardens in the Landscape of Western Warsaw] *Problemy Ekologii Kajobrazu* 22: 139–50.

Mason, C.W., C.R. Stoof, B.K. Richards, S. Das, C.L. Goodale, and T.S. Steenhuis. 2017. Hotspots of Nitrous Oxide Emission in Fertilized and Unfertilized Perennial Grasses on Wetness-Prone Marginal Land in New York State. *Soil Science Society of America Journal* 81: 450–8.

McBride, M.B., H.A. Shayler, H.M. Spliethoff, R.G. Mitchell, L.G. Marquez-Bravo, G.S. Ferenz, J.M. Russell-Anelli, L. Casey, and S. Bachman. 2014. Concentrations of Lead, Cadmium and Barium in Urban Garden-Grown Vegetables: The Impact of Soil Variables. *Environmental Pollution* 194: 254–61.

McKenna-Neuman, C.L. 2011. Dust Resuspension and Chemical Mass Transport from Soil to the Atmosphere. In, *Handbook of Chemical Mass Transport in the Environment*, edited by L.J. Thibodeux, and D. Mackay, 453–93. Boca Raton, FL: CRC Press.

Méndez, V.E., C.M. Bacon, and R. Cohen. 2016. Introduction: Agroecology as a Transdisciplinary, Participatory, and Action-Oriented Approach. In, *Agroecology. A Transdisciplinary, Participatory and Action-Oriented Approach*, edited by V.E. Méndez, C.M. Bacon, R. Cohen, and S.R. Gliessman, 1–22. Boca Raton, FL: CRC Press.

Menefee, D.S., and G.M. Hettiarachchi. 2018. Contaminants in Urban Soils: Bioavailability and Transfer. In, *Urban Soils*, edited by R. Lal, and B.A. Stewart, 175–98. Boca Raton, FL: CRC Press.

Meuser, H. 2010. *Contaminated Urban Soils*. Dordrecht: Springer.

Mies, M., and V. Bennholdt-Thomsen. 1999. *The Subsistence Perspective: Beyond the Globalized Economy*. London: Zed Books.

Mitchell, R.G., H.M. Spliethoff, L.N. Ribaudo, D.M. Lopp, H.A. Shayler, L.G. Marquez-Bravo, V.T. Lambert, G.S. Ferenz, J.M. Russell-Anelli, E.B. Stone, and M.B. McBride. 2014. Lead (Pb) and Other Metals in New York City Community Garden Soils: Factors Influencing Contaminant Distributions. *Environmental Pollution* 187: 162–9.

Mok, H.-F., V.G. Williamson, J.R. Grove, K. Burry, S.F. Barker, and A.J. Hamilton. 2014. Strawberry Fields Forever? Urban Agriculture in Developed Countries: A Review. *Agronomy for Sustainable Development* 34: 21–43.

Moseley, W.G. 2001. African Evidence on the Relation of Poverty, Time Preference and the Environment. *Ecological Economics* 38(3): 317–26.

Mougeot, L.J.A. 2015. Urban Agriculture in Cities in the Global South: Four Logics of Integration. In, *Food and the City, Histories of Culture and Cultivation*, edited by D. Imbert, 163–93. Washington, DC: Dumbarton Oaks.

Mudu, P., and A. Marini. 2018. Radical Urban Horticulture for Food Autonomy: Beyond the Community Gardens Experience. *Antipode* 50(2): 549–73.

Nabulo, G., S.D. Young, and C.R. Black. 2010. Assessing Risk to Human Health from Tropical Leafy Vegetables Grown on Contaminated Urban Soils. *Science of the Total Environment* 408: 5338–51.

Nyantakyi-Frimpong, H., G. Arku, and D.K. Baah Inkoom. 2016. Urban Agriculture and Political Ecology of Health in Municipal Ashaiman, Ghana. *Geoforum* 72: 38–48.

O'Connor, J. 1988. Capitalism, Nature, Socialism. A Theoretical Introduction. *Capitalism Nature Socialism* 1(1): 11–38.

Orsini, F., R. Kahane, R. Nono-Womdin, and G. Giaquinto. 2013. Urban Agriculture in the Developing World: A Review. *Agronomy for Sustainable Development* 33: 695–720.

Paltseva, A., Z. Cheng, M. Deeb, P.M. Groffman, R.K. Shaw, and M. Maddaloni. 2018. Accumulation of Arsenic and Lead in Garden-Grown Vegetables: Factors and Mitigation Strategies. *Science of the Total Environment* 640–1: 273–83.

Parece, T.E., and J.B. Campbell. 2017a. A Survey of Urban Community Gardeners in the USA. In, *Global Urban Agriculture: Convergence of Theory and Practice Between North and South*, edited by A.M.G.A. WinklerPrins, 38–49. Boston, MA: CABI.

Parece, T.E., and J.B. Cambpell. 2017b. Assessing Urban Community Gardens' Impact on Net Primary Production Using NDVI. *Urban Agriculture & Regional Food Systems* 2: 1–17.

Parisio, S. 2018. *Sustainable Management of Urban Soil & Fill Material*. Paper Presented at Federation of New York Solid Waste Associations Solid Waste & Recycling Conference. The Sagamore on Lake George, Bolton Landing, NY, 21 May.

Pataki, D.E., M.M. Carreiro, J. Cherrier, N.E. Grulke, V. Jennings, S. Pincetl, R.V. Pouyat, T.H. Whitlow, and W.C. Zipperer. 2011. Coupling Biogeochemical Cycles in Urban

Environments: Ecosystem Services, Green Solutions, and Misconceptions. *Frontiers in Ecology and the Environment* 9(1): 27–36.

Pauleit, S., and J.H. Breuste. 2011. Land-Use and surface-Cover as Urban Ecological Indicators. In, *Urban Ecology. Patterns, Processes, and Applications*, edited by J. Niemelä, 19–30. Oxford: Oxford University Press.

Platt, S.A. 2014. Death by Arugula: How Soil Contamination Stunts Urban Agriculture, and What the Law Should Do About It. *Minnesota Law Review* 97: 1507–48.

Ponce, Marianna, and Lucrecia Donoso. 2009. Urban Agriculture as a Strategy to Promote Equality of Opportunities and Rights for Men and Women in Rosario, Argentina. In, *Women Feeding Cities: Mainstreaming Gender in Urban Agriculture and Food Security*, edited by A. Hovorka, H. de Zeeuw, and M. Njenga, 157–66. Warwickshire: Practical Action Publishing.

Posey, D.A., and W. Balée, eds. 1989. *Resource Management in Amazonia: Indigenous Folk Strategies*. Bronx: New York botanical Garden.

Pouyat, R.V., K. Szlavecz, I.D. Yesilonis, P.M. Groffman, and K. Schwarz. 2010. Chemical, Physical and Biological Characteristics of Urban Soils. In, *Urban Ecosystem Ecology*, edited by J. Aitkenhead-Peterson, and A. Volder, 119–52. Agronomy Monograph 55. Madison WI: American Society of Agronomy, Crop Science Society of America, Soil Science Society of America.

Prain, G., and D. Lee-Smith. 2010. Urban Agriculture in Africa: What Has Been Learned? In, *African Urban Harvest: Agriculture in the Cities of Cameroon, Kenya and Uganda*, edited by G. Prain, N. Karanja, and D. Lee-Smith, 13–35. New York: Springer.

Reynolds, K., and N. Cohen, eds. 2016. *Beyond the Kale. Urban Agriculture and Social Justice Activism in New York City*. Athens: University of Georgia Press.

Richards, P. 1985. *Indigenous Agricultural Revolution: Ecology and Food Production in West Africa*. London: Hutchinson.

Roitman, S., and M. Bifarello. 2010. Urban Agriculture and Social Inclusion in Rosario, Argentina. *Inclusive Cities Observatory*. www.uclg-cisdp.org/sites/default/files/Rosario_2010_en_final.pdf (Accessed 23 July 2020).

RUAF. 2017. Urban Agroecology. *Urban Agriculture Magazine* 33. https://ruaf.org/assets/2019/11/Urban-Agriculture-Magazine-no.-33-Urban-Agroecology.pdf (Accessed 12 July 2020).

Sarzhanov, D.A., V.I. Vasenev, I.I. Vasenev, Y.L. Sotnikov, O.V. Ryzhkov, and T. Morin. 2017. Carbon Stocks and CO_2 Emissions of Urban and Natural Soils in Central Chernozemic Region of Russia. *Catena* 158: 131–40.

Säumel, I., I. Kotsyuk, M. Hölscher, C. Lenkereit, F. Weber, and I. Kowarik. 2012. How Healthy Is Urban Horticulture in High Traffic Areas? Trace Metal Concentrations in Vegetable Crops from Plantings within Inner City Neighbourhoods in Berlin, Germany. *Environmental Pollution* 165: 124–32.

Sawio, C.J. 1998. Managing Urban Agriculture in Dar es Salaam. https://idl-bnc-idrc.dspacedirect.org/bitstream/handle/10625/23305/108517.pdf?sequence=1 (Accessed 12 September 2018).

Schreck, E., Y. Foucault, G. Sarret, S. Sobanska, L. Cécillon, M. Castrec-Rouelle, G. Uzu, and C. Dumat. 2012. Metal and Metalloid Foliar Uptake by Various Plant Species Exposed to Atmospheric Industrial Fallout: Mechanisms Involved for Lead. *The Science of the Total Environment* 427–8: 253–62.

Schwarz, K., B.B. Cutts, J.K. London, and M.L. Cadenasso. 2016. Growing Gardens in Shrinking Cities: A Solution to the Soil Lead Problem? *Sustainability* 8: 141–52.

Silva, E., and A. Pfeiffer. 2016. Agroecology of Urban Farming. In, *Cities of Farmers. Urban Agricultural Practices and Processes*, edited by J.C. Dawson, and A. Morales, 107–25. Iowa City: University of Iowa Press.

Smith, N. 1984. *Uneven Development: Nature, Capital, and the Production of Space*. New York: Blackwell.

Tian, H., G. Chen, C. Lu, X. Xu, W. Ren, B. Zhang, K. Banger, B. Tao, S. Pan, M. Liu, C. Zhang, L. Bruhwiler, and S. Wofsy. 2015. Global Methane and Nitrous Oxide Emissions from Terrestrial Ecosystems Due to Multiple Environmental Changes. *Ecosystem Health and Sustainability* 1(1): 4. http://dx.doi.org/10.1890/EHS14-0015.1.

Tracey, D. 2011. *Urban Agriculture. Ideas and Designs for the New Food Revolution*. Gabriola Island: New Society Publishers.

van der Ent, Anthony, Guillaume Echevarria, Alan J.M. Baker, and Jean Louis Morel, eds. 2018. *Agromining: Farming for Metals*. Berlin: Springer International Publishing.

Vasenev, V.I., J.J. Stoorvogel, R. Leemans, R. Valentini, and R.A. Hajiaghayeva. 2018. Projection of Urban Expansion and Related Changes in Soil Carbon Stocks in the Moscow Region. *Journal of Cleaner Production* 170: 902–14.

Viehweger, K. 2014. How Plants Cope with Heavy Metals. *Botanical Studies* 55: 35. www.as-botanicalstudies.com/content/55/1/35 (Accessed 2 August 2018).

Viljoen, A., and K. Bohn, eds. 2014. *Second Nature Urban Agriculture: Designing Productive Cities*. New York: Routledge.

Voigt, A., and T.E. Leitão. 2016. Lessons Learned. Indicators and Good Practice for an Environmentally-Friendly Urban Garden. In, *Urban Allotment Gardens in Europe*, edited by S. Bell, R. Fox-Kämper, N. Keshavarz, M. Benson, S. Caputo, S. Noori, and A. Voigt, 165–97. London: Routledge and Earthscan.

von Hoffen, L.P., and I. Säumel, I. 2014. Orchards for Edible Cities: Cadmium and Lead Content in Nuts, Berries, Pome and Stone Fruits Harvested Within the Inner City Neighbourhoods in Berlin, Germany. *Ecotoxicology & Environmental Safety* 101: 233–9.

Warming, M., M.G. Hansen, P.E. Holm, J. Magid, T.H. Hansen, and S. Trapp. 2015. Does Intake of Trace Elements Through Urban Gardening in Copenhagen Pose a Risk to Human Health? *Environmental Pollution* 202: 17–23.

Wiseman, C.L.S., F. Zereini, and W. Püttmann. 2015. Metal and Metalloid Accumulation in Cultivated Urban Soils: A Medium-Term Study of Trends in Toronto, Canada. *Science of the Total Environment* 538: 564–72.

Witzling, L., M. Wander, and E. Phillips. 2011. Testing and Educating on Urban Soil Lead: A Case of Chicago Community Gardens. *Journal of Agriculture, Food Systems, and Community Development* 1(2): 167–85.

Wortman, S.E., and S.T. Lovell. 2013. Environmental Challenges Threatening the Growth of Urban Agriculture in the United States. *Journal of Environmental Quality* 42: 1283–94.

Xiong, T., T. Leveque, M. Shahid, Y. Foucault, S. Mombo, and C. Dumat. 2014. Lead and Cadmium Phytoavailability and Human Bioaccessibility for Vegetables Exposed to Soil or Atmospheric Pollution by Process Ultrafine Particles. *Journal of Environmental Quality* 43(5): 1593–600.

Yadav, P. 2010. Heavy Metals Toxicity in Plants: An Overview on the Role of Glutathione and Phytochelatins in Heavy Metal Stress Tolerance of Plants. *South African Journal of Botany* 76(2): 167–79.

Yadav, P., K. Duckworth, and P.S. Grewal. 2012. Habitat Structure Influences Below Ground Biocontrol Services: A Comparison Between Urban Gardens and Vacant Lots. *Landscape and Urban Planning* 104: 238–44.

7

LOCAL CONTINGENCIES OF URBAN CULTIVATION

Now that the biophysical transformations involved in urban food production and the political importance of studying and understanding biophysical processes have been clarified, the dialectical relationship among ecosocial processes (i.e. biophysical and social) can be more effectively assessed in specific contexts to illustrate what holds promise relative to urban cultivation's role in promoting the construction of an alternative, ecosocialist order. The biophysical and social effects of and challenges to urban food production are decisively dependent on local conditions as well as what happens more widely. What people do affects and is affected by the rest of an urban ecosystem. To illustrate, urban cultivation area parameters and food production practices, which affect biodiversity, element cycling, and other biophysical processes, are fundamentally shaped by cultivators' social circumstances, such as income level and available free time (Loram et al. 2008). The circumstances are determined by multiple kinds of relations of power or, to restate it more directly, multiple forms of oppression. Biophysical transformations of cities through urban cultivation therefore lead to multiple directions that are good for some species over others and, within society, for some people compared to others. At the same time, there are biophysical dynamics independent of and affected by past or current human impacts that pose obstacles to or provide favourable conditions for urban cultivation. To get a fuller idea, one should be mindful of the interactions at the urban (and wider) ecosystem scale, the biophysical dynamics within cultivated spaces compared to other kinds of green spaces and built-up areas, and the social relations that give rise to and maintain these different urban spaces.

There is a great variety of contexts and food production systems, combinations of factors, and possible permutations in the order of factors' importance. We draw from examples that may broadly represent the diversity of biophysical and social conditions in terms of biome-climate region, country situation, and city size. However, the primary aim is to show how food production intertwines with

DOI: 10.4324/9781003131281-7

other biophysical and social processes to produce different kinds of situations that beckon the development of place-specific projects and solutions. Politically, this to us implies a bottom-up confederated communal structure, like the current Zapatista and Rojava examples, as the most practical way of organising cities, if one is keen on ecologically sustainable egalitarianism leading to a class-free society. With the focus here being on urban food production, the main split is also generally about the degree of producers' livelihood dependence on urban food production. In most cities, food production is more akin to smallholder farming (Mougeot 2015; Orsini et al. 2013). It is practised in ways that are not markedly different from the countryside, and it can also involve livestock. In other words, unlike the sort of food production that goes on in places like New York City, Beijing, or Berlin, many people in most cities worldwide depend on food production for their livelihoods, and productivity levels may even go beyond meeting household needs. These forms of urban food production are scarcely comparable to urban gardens in capital-wealthy countries, which scandalously overwhelm the literature.

At the same time, because we are concerned with the specifically ecosocialist potentials of urban food production, it may very well be that urban farming in Cuba and urban community gardens elsewhere merit disproportionate attentiveness. This is because such projects run counter to capitalist prerogatives (Certomà and Tornaghi 2015). For example, even beyond Cuba, problems with domestic water shortages in Zimbabwe have promoted the development of community gardens that furnish year-round water as well as gardening training and tools (Moses 2015). This is not to suggest that such, sometimes avowedly communalistic, food-producing areas are more important than other similar activities not carried out with explicit political intent. Urban cultivation that resembles subsistence-oriented smallholder farming is just as politically significant insofar as it demonstrates that food production based on non-capitalist principles not only exists, but also succeeds in feeding people. Put differently, conditions already exist for a transition to ecosocialist forms of food production within and beyond cities (what this could mean is discussed in the final chapter).

The viability of urban cultivation, however, rests on favourable social as well as biophysical relations. Perhaps to state the obvious, ensuring that food can be produced and is safe to consume necessitates sensitivity to biophysical contingencies. The necessity is also political in terms of reconciling environmental and social concerns or, even better, unifying technical biophysical with social justice understandings and priorities. However, the vast majority of studies on urban food production are hindered by an inveterate institutional compartmentalisation of fields of knowledge that pre-empts holistic and relational understanding. What little has been done that deals with both society and the rest of nature does so in an additive way, first describing one and then the other, but never really bringing the sets of relations together, and certainly not with any explicit or clear political commitments (e.g. Bell et al. 2016). The difficulty is compounded by a highly uneven geographical and thematic coverage of urban cultivation. Yet, one can still gather together different studies to provide some general ecosocial understanding of urban cultivation that can help delineate longer-term political possibilities. We discuss five such case studies here.

Mixed market and subsistence production in Tamale

Tamale, the capital of Ghana's Northern Region and a historical trade-route centre, is a rapidly expanding city of some 370,000 people, 433 km north of Accra, the capital. Located in a savanna plain, Tamale usually receives most rainfall in summer to mid-fall. As the city is growing, urban cultivation space is diminishing. Climate characteristics feature cooler weather in late fall to mid-winter followed by a drier Harmattan (windy) season. Irrigation water comes from local reservoirs and small channels (gutters, standpipes). Freshwater supply is constrained by a low water table and a few streams. Gathering and storing rainwater during the rainy season is common. Maize, yam, and rice are staple crops locally, mainly grown in the city's outskirts (Bellwood-Howard et al. 2018; Karg et al. 2016).

Recent climate change has modified seasonal patterns of precipitation while urban expansion has been encroaching on surrounding farmland. Farming households, including those in the countryside, have responded by introducing new cultivars, diversifying the ways they gain their livelihood, and intensifying crop production over smaller patches of land (Gyasi et al. 2014). Maize is usually combined with leafy vegetables, including jute mallow, lettuce, and cabbage. Crops like jute mallow and amaranth can be harvested multiple times, but the Harmattan season makes growing brassica varieties like cabbage more feasible (Bellwood-Howard et al. 2018). Many of the vegetables grown and sold within the city are produced using agrochemicals. Crucially, Tamale's inhabitants' nutritional needs are often unmet in general because of lack of effective food access (due to high poverty rates). Nearly a third of the children are malnourished and are consequently hit with lifelong health problems. The amount of food produced within the city is nowhere near any possibility of meeting the basic food needs for Tamale's dwellers (Karg et al. 2016).

A regular regional climate cycle like the Harmattan enables women marketers in Tamale to fetch more profit from selling jute mallow, in addition to other produce (Bellwood-Howard and Bogweh Nchanji 2017). This presumes a social arrangement where vegetables are turned into commodities and where there are large inequalities in production technologies, but the specificity of what is grown and sold is also shaped by environmental factors like Harmattan timing and intensity. Or, put differently, some people are better placed to use existing relations of power cloaked in crop growth seasonality (i.e. a technical issue) for relative social advantage. This may happen when women marketers will be better positioned during the Harmattan season to make greater claims over a vegetable harvest made by groups of other women (ibid., 85). In this instance, crop selection is as much a function of the local climate as of local cultural histories inflecting culinary preferences and the absence of greenhouses or row covers. With regional climate change, matters may also shift socially, but not because of environmental change alone. A shortened Harmattan season could only prompt other ways to justify the distinction and economic differentials among food producer, marketer, and retailer. Environmental conditions obviously do not determine social relations, but the latter rely on the former for their very existence as well as for ideological support to maintain certain social arrangements that may or may not be associated with social inequalities.

At the same time, crop selection is not only the product of local cultural histories or quirks. Over the past decade or so, cabbage cultivation has been expanded alongside that of other vegetables because they have become more lucrative. This shift is made possible by greater integration into capitalist markets and the commodification of vegetable production, which favours some people over others. Such greater profitability is coming at the expense of public health, however. As mentioned earlier, vegetables, which are often eaten raw, are being irrigated with waste- and grey-water and grown under heavy doses of pesticides (Bogweh Nchanji et al. 2017). This is related to water shortages pitting different forms of farming against each other and against other uses, such as industrial processing and residential needs (Bellwood-Howard and Bogweh Nchanji 2017, 86–7). A healthcare disaster waiting to happen, water scarcities will worsen an already disastrous public health situation.

These dynamics all have political consequence because it should always be made clear that what is ecologically sustainable does not necessarily translate into particular social outcomes. What happens within a society largely depends on how people within that society relate to each other (power relations) and relative to how environments and other species change. Climate change, which looms very large in the case of Tamale, and related public health problems and threats from a history of immiseration (more than half of Tamale inhabitants live in poverty), and vegetable contamination will likely heighten existing inequalities in food provisioning. Such inequalities need to be tackled especially along gendered class lines (e.g. land tenure arrangements, which presently favour men) and alongside bringing relief from economic pressure over produce sales. Delinking from the world capitalist economy can be helpful, as the late Samir Amin (1973) long advocated. The widespread poverty and horrific levels of privation are the consequence of "structural adjustment" impositions from foreign powers, often under the aegis of the International Monetary Fund. Thus, Ghana's general situation is a consequence of neo-colonial relations, mainly with the US and the UK (Langan 2017; Nkrumah 1965). Yet, local relations of power also need to be addressed so that everyone can have equal land and food access or otherwise participate meaningfully in the local agro-food economy.

Conventional forms of farming in rapidly expanding Dar es Salam

Dar es Salaam, a major coastal port city and former capital of Tanzania, is among the most rapidly expanding cities in the world, much more so than Tamale. Urban agriculture researchers have showered a lot of attention on that city, especially in the 1990s. This is because of the substantial economic role played by food production (both for sale and subsistence) and the relatively high productivity in leaf vegetables, which, in the 1990s, provided a staggering 90 per cent of the total consumed (Dongus and Nyika 2000). However, staple foods like maize, potatoes, and rice mainly come from distant, largely smallholder farms in the countryside (Wegerif and Wiskerke 2017).

The city, so far, is spread out over ca. 1,400 km^2. Of this area, about 650 km^2 (46 per cent of the urban area) was under cultivation by the 1990s, and recently put under encroachment pressures to be built on (even to make golf courses, on occasion). Sand and tidal swamps give way to a coastal plain underlain by limestone, followed to the west by the Pugu hills (from 100 to 300 m above sea level), characterised by highly weathered slopes and overlain by unconsolidated gravelly clay deposits and clayey soils. The plain is crossed by four major rivers that end up as creeks and mangrove swamps prior to entering the Indian Ocean. These alluvial plains are surrounded by silt-clay soils containing high amounts of organic matter. The water table tends to be close to the surface as a result of abundant water supply, the presence of several rivers, and proximity to the ocean (Justin et al. 2018; Mtoni et al. 2013).

Local ecosystems vary among forest (some of them remnant), grassland (including golf courses), riparian, dune (coastal), and estuarine, with much built-up, paved area, as well as some extractive industries (e.g. quarries). Only 5.1 per cent of the city is permanently covered by vegetation, but there are large parks along coastal areas and riverbanks. A little more than half of the vegetation is composed of bushland and nearly a quarter is riparian, followed by marshes, mangroves, and mixed forest. Rains arrive in spring and fall, for the most part (Tibesigwa et al. 2018).

Located close to the Equator, the city is within a tropical climate that ensures high year-round humidity and temperatures, made quite oppressive by a recently intensifying urban heat island effect (Ndetto and Matzarakis 2013), and plenty of rainfall (an average of more than 1,000 mm per year). Climate change is affecting the city with increasing average temperatures, more frequent and severe droughts, sea-level rise, and more concentrated rains, the latter two effects raising the frequency and magnitude of floods. Flooding has been made even more destructive by expanding paved area and poor preventive infrastructure. There is with this a potential for expanding the breeding ground for the *Anopheles* mosquitoes (a malaria vector), given that they appear to survive well enough in many kinds of stagnant water habitats, regardless of pollution or turbidity levels (Bisset et al. 2011; Sattler et al. 2005). Sea-level rise is also contributing to a quickening of coastal erosion as well as saltwater intrusion into groundwater supplies, which degrades the quality of drinking and irrigation water (Garcia-Aristizabal et al. 2015; Justin et al. 2018; Kiunsi 2013; Sappa et al. 2015).

Thanks in part to the demise of the Ujamaa system (1967–1985), which had held urbanisation in check, there has been much migration into the city from the countryside, contributing, alongside incoming international migrants, to a total population surpassing 4 million (Coulson 2013; Tripp 1997). But new construction and roadways have not been linked only to the influx of migrants. Transport infrastructure has emerged as well in relation to tourism, land speculation, and trade in raw materials and finance, some related to violent and genocidal extractive industries in the eastern regions of the neighbouring Democratic Republic of Congo. Between 1991 and 2009, built-up area increased by 59 per cent (from 130 to 206 km^2) and cultivated land also expanded by 15 per cent (from 737 to 848 km^2), all largely at the expense of forests (including mangroves) and bushland. The more central parts of the city have been concomitantly subjected to rapid industrialisation in infrastructure, with commuter railway lines, increasing motorised traffic, and high-rise buildings being among the recent additions to local daily life. Green spaces, consequently, have been increasingly destroyed to make room for all sorts of built-up areas, from shacks and dirt paths to posh residences, office spaces, and paved roadways. Much of the expansion has been at the urban edges, where basic infrastructure is often lacking and untreated sewage, wastewater, and industrial waste at times leads to outbreaks of water-borne diseases. However, soil and water pollution problems arise as well from regular, everyday waste disposal, industrial and vehicular emissions, and farm contaminant releases, mainly organic wastes and agrochemicals (Briggs and Mwamfupe 1999; Himberg 2016; Mahugija, Kayombo, and Peter 2017; Mbuligwe and Kassenga 1997; Mkalawa 2016; Mlozi 1997; Mtoni et al. 2013; Mwegoha and Kihampa 2010; Sappa et al. 2015; Sawio 1998, 16; Tibesigwa et al. 2018).

Food production has been an important source of income as well as subsistence especially among the poor, who may be farmers or farmworkers in small farming operations. Government inducements strengthened urban cultivation by making more space available to grow food in the early 1970s, during an economic emergency related to the 1973 oil crisis (Coulson 2013; Tripp 1997). Since that time, urban cultivation has contracted and then expanded again with another, longer-lasting economic downturn (Nugent 2000, 71). This time, many became impoverished and took up farming to stave off hunger, while many displaced rural people moved to the city.

Taking up about 46 per cent of the urban area, farming has been expanding moderately, as mentioned earlier. A third or more of cultivated land is occupied by mixed farming, while the rest involves conventional cropping systems and vegetable gardens. The municipality has long recognised and allowed urban food growing. Livestock-raising has also been permitted, so that dairy cattle, goats, sheep, pigs, and fowl (especially chickens) are kept by many households. In fact, cattle were freely roaming in residential areas in the 1990s, and chicken manure is often used as fertiliser for vegetable plots (Nugent 2000, 85–6). Vegetable production and livestock pasturing have also been known to be practised over polluted areas, such as along the Msimbazi River. Use of river water contaminated by cadmium, lead, and zinc has also been reported (Mwegoha and Kihampa 2010; Sawio 1998, 12–13).

Cultivation presently occurs in many parts of the city, from tiny backyards to fields of several hectares. Farming is relatively productive and profitable, with possibilities as well for growing staples (Sawio 1998; Tibesigwa et al. 2018). An estimate from the 1990s posits that 90 per cent of leafy greens and 60 per cent of milk consumed in the city are produced by local open field and backyard cultivators and cattle keepers (Armar-Klemesu 2000, 104; Nugent 2000, 81). This is mostly the case for farming in open spaces, which are often public and available. Crop production in such places is largely market-oriented and predominantly run by men (Nugent 2000, 80). The land, however, may be susceptible to appropriation by other kinds of businesses, especially in cases of insecurity in land access or ownership. This is even if tenure rights may be had by squatters who stay on public land for a decade or so (Tibesigwa et al. 2018). Hence, urban farmer displacement may increase as speculators and other businesses seek to capitalise on the urban expansion trend, with the local government proving ineffective at curbing such speculation (Drechsel and Dongus 2010). Women have often been able to gain more economic independence by engaging in farming business but usually through informal sales or by selling surplus from subsistence plots (Nugent 2000, 81; Tripp 1997, 13).

Biophysical changes may be even more menacing than economic pressures, though. Without adequate state support (Mwalukasa 2000), farmers may find it increasingly difficult to produce enough to maintain profitability in the face of a spreading pollution problem and more frequent disasters associated with climate change. But this may yet be the least of the problems. The fact that most of the consumed leafy greens, which tend to be trace element accumulators, are locally produced should provoke thorough investigation, given the reported problems of irrigation water and soil pollution with heavy metals. This is a potentially explosive health matter that appears not to have been met with any institutional action so far. The potential hazard adds to known produce, meat, and water contamination directly caused by urban farmers' pesticide use (Mahugija, Khamis, and Lugwisha 2017; Mahugija, Chibura, and Lugwisha 2018). This deleterious impact of urban farming adds to wider health problems and shortened lifespans due to periodic extreme weather events (floods, heat waves, etc.), resulting from the combined effects of global warming and local impacts (e.g. expansion of paved area), and disease outbreaks related to waste disposal and stagnant water (McCrickard et al. 2017). Institutional support for farming is lacking where farmers use polluted land or irrigation water. The spread of contamination problems seems to be undermining the market-oriented cultivators rather than subsistence-oriented ones (Drechsel and Dongus 2010), aside from the potential pollutant exposure risk increasing for thousands of urban cultivators.

Overall, urban cultivation and livestock-raising attain substantial productivity levels as a sector integrated into the local economy. Commercial farming over large open spaces is mostly held by men, while subsistence production is mainly done by women. The market orientation of urban food growing also heightens class differentiation and seems to promote further marginalisation of poorer urban dwellers. Prospects for a sustainable food self-sufficient city are not as rosy as once thought in the 1990s by some non-profit international consultants, scholars, and policy makers (see, e.g. Howorth, Convery, and O'Keefe 2001; Lee-Smith 2010; Schmidt 2012). To some extent, the ultimate exclusion of urban food growing from the master plan of the city government (Mkwela 2013) speaks for itself in terms of where priorities currently lie. Urban farming seems increasingly dependent on whether food produced within the city will fetch profits comparable to other kinds of business. Commercial disadvantages and ever-increasing construction pressures are more likely to efface pastures and farming fields. This can only be exacerbated by capitalist-friendly land tenure policies (Bersaglio and Kepe 2014). Perhaps even more worrisome, and flying in the face of urban sustainability claims, is the increasing intensity of challenges brought by combinations of global atmospheric and local biophysical changes (e.g. environmental contamination, increasing pathogen outbursts). Importantly, the latter is caused in part by urban farming itself. This has long been recognised by a few and mostly with regard to cattle (e.g. Mlozi 1996), but without any constructive proposals to determine the extent and form of the problems, to introduce measures to safeguard farmworkers, ensure health safety in farm products, and establish social justice policies to promote egalitarian access to uncontaminated and nutritious food. The deleterious effects bubbling up from the bowels of accelerated capitalist urbanisation will doubtless be meted out most to the most marginalised, whose dependence on urban-produced food increases when food prices shoot through the roof, but this outcome seems to be largely omitted in the copious scholarship on Dar es Salaam.

Gardening as struggle against landlordism in Rome

In contrast to the global South cities, Italy's urban cultivation is a relative reintroduction that only marginally contributes to food self-provisioning, although it has been an important complement to rural food procurement during emergencies (like war time). Rome, the capital, is of special interest here because it has witnessed a sudden popularity and spread of food production by way of community gardens, in contrast to past urban food production forms (Attili 2013; Mudu and Marini 2018). The city has developed and expanded over a couple of millennia in a Mediterranean climate that is becoming rainier, and over a hilly area dissected by the partly canalised and meandering River Tiber. Rome's floodplain is partly surrounded by seven hills reaching little more than 100 m in elevation, which are remains of ancient volcanoes. The local water supply is in part fed by underground aquifers within limestone rock layers. Some of these layers are made of marble used profusely in antiquity to produce the built environment that exists, in various states of ruin, to this day. Other rock layers are composed of travertine, which contains arsenic and has been a subject of concern over water contamination. Another source of contaminants is in the volcanic sediments (pozzolanic ash in particular) that contain high amounts of lead (Calace et al. 2012; Ventriglia 2002). There are, in other words, long pre-existing geogenic sources of trace element contamination that need to be accounted for when assessing urban cultivation.

The long-term result of urban construction is a patchwork of large green spaces, especially towards the periphery, and paved area in the historical centre. In fact, there is much farmland immediately surrounding the city. However, only a small fraction of Rome (ca. 5 per cent) derives from the various construction periods of antiquity. For the most part, Rome is actually a new city, built largely since the 1950s (Mudu and Marini 2018, 7). The result of recent urbanisation, with industrialised means of consumption and transportation, is evident throughout. Large parks towards the urban centre contain high quantities of heavy metal trace elements near roadways (such as cadmium, mercury, zinc, and increasingly platinum), as well as moderate contamination from PCBs, PAHs, and organochlorines. These pollutants are likely due to vehicular emissions and a liberal use of pesticides, among other impacts (Angelone, Corrado, and Dowgiallo 1995; Cenci et al. 2008; Cinti et al. 2002).

Imprints of recent land uses have contributed to an under-surface made of various geological layers overlapping with building debris and foundations, as well as mixtures of human-transported sediments and assorted rubbish. Many soils have thus developed out of anthropogenic materials originating from a wide range of intensive land uses over the recent past, including alkaline, calcium-rich construction debris (Della Seta and Della Seta 1988). As would be expected from this, soils tend to be relatively alkaline, which means that plant roots will not absorb most heavy metal trace elements easily (although particles may become stuck on the tissue surface, especially on roots). This is an important consideration, given the combined geogenic and anthropogenic sources of contaminants. Even though Rome has been largely spared the polluting excesses of full-fledged manufacturing sectors (Brunori and Di Iacovo 2014), most inhabitants, as in many Italian cities, nevertheless suffer from chronic air pollution exposure that recently seems to be worsening, with recurring standards-smashing episodes lasting days to weeks. Otherwise, most contamination is localised and emanates from nearby processing industries and service sector activities, or their legacies (Calace et al. 2012; Conio and Porro 2004; Fusco et al. 2001).

Urban food producers usually confront dry summers and rainy winters. Vegetables require irrigation for at least part of the year, but production, with a judicious selection of crops, can be year-long. Animal husbandry is virtually banned, except towards the peri-urban fringes, so its products are at most of marginal relevance. Vegetable production, largely geared to local tastes (tomatoes, various brassicas, beans, etc.), is also very low and largely insufficient to satisfy the yearly needs of even single households. However, the produce is integrated with locally served dishes (offered at gatherings or even sold in social centres or squats), and it is consumed by gardeners in as much quantity as can be had. The already exiguous productivity level is jeopardised by recent environmental changes and long-standing pollution problems. With climate change, there are increasingly extreme weather events, punctuated by long periods of drought or sudden and intense precipitation. Over the past few decades, sudden bursts of intense rainfall seem increasingly frequent in summer, raising the likelihood of flash floods and wider pollutant dispersal.

Urban gardens are situated in an array of combined and uneven contamination processes that, for the most, part should not affect vegetables through soil nutrient root absorption, given the prevailing alkalinity. But the problem seems to lie above gardens, by way of falling dust, or from within, by way of soil particles kicked up and re-deposited on vegetables. One of us (Engel-Di Mauro 2018) had the opportunity to learn about and contribute to a contamination study of two community gardens in Rome that are the result of direct struggles to convert urban spaces into self-managed areas. Both sites are underlain by sediment derived from lead-enriched grey pozzolanic ash (GsF 2012; Ventriglia 2002). No agrochemicals are used, which is to demonstrate an alternative to conventional food production. Local piped water is used for irrigation and on-site harvest processing.

One of the community gardens is located within Forte Prenestino, a large late nineteenth-century military fortress where a squat was established in 1986 (Figures 7.1 and 7.2). The fortress had been made part of a public park by the late 1970s, but it has not been put under any specific use. Forte Prenestino is run as a set of collectives, and periodic assemblies are called to make decisions over the running of the squat and about ongoing and new projects. The garden within the squatted fortress was not set up until 2011, in a fenced area of roughly 200 m^2. The garden is surrounded in part by trees and part of the fortress walls, so it is relatively sheltered. It is managed by a small collective within the larger squatters' collective. The

FIGURE 7.1 The Forte Prenestino squat's vegetable garden, Rome (photo by Salvatore Engel-Di Mauro 2014)

area is divided into raised beds made from excavated material from adjacent soil. Cropping areas are permanently covered by imported straw and managed according to synergistic principles (Hazelip 2014). These entail mulching, crop residue recycling, and avoidance of soil reworking (e.g. digging, tillage) and inputs (e.g. fertiliser, insecticide).

The other community garden, Orto Insorto (Insurgent Garden), was a pre-emptive action by neighbourhood collectives in 2011 to prevent the potential for

FIGURE 7.2 Tomatoes grown according to synergistic principles in the garden within Forte Prenestino squat, Rome (photo by Salvatore Engel-Di Mauro 2014)

land speculation over an abandoned lot of little more than two hectares in area. The garden, established the same year of the squatting action, is but a small portion of the total area, which is fenced and adjacent to a main road. The entire area sits on a buried waste dump now 25–55 cm below the surface. There are open-air meeting spaces, grassy areas, fruit trees, directly cropped areas, and several plastic containers filled with imported "organic" soil (i.e. officially devoid of contaminants or agro-chemical treatments). The site is within 100 m from a construction material plant specialising in industrial-grade paints, thermo-hydraulics, and building material recovery. The garden areas are located more than 50 m from the plant but within 30 m of the adjacent road (Figure 7.3).

A preliminary assessment of the contamination problem was done by sampling for air deposition, gardening inputs (e.g. straw, imported soil, irrigation water), vegetables, and the soils on which the vegetables were grown. An important example of environmental monitoring from below, the Orto Insorto collective had already conducted a study and had found high levels of lead. The levels varied between 552 and 3,170 ppm lead within the top 10 cm. For those unfamiliar with soil quality standards, 60 ppm lead in soil is already pushing the official limits of acceptability (at least for the US Environmental Protection Agency), especially for growing vegetables. However, the amounts of lead in the vegetables were comfortably below unhealthy levels of 0.1 ppm, at least according to European Union

FIGURE 7.3 The Orto Insorto gardening and recreational areas, Rome (photo by Salvatore Engel-Di Mauro 2014)

norms, except in the case of lettuce (0.47 ppm). The reason for having to rely on different institutions for various safety standards is because there exists no globally unified set of guidelines to interpret test results for trace elements (for some trace elements, there are no regulatory standards at all). In any event, the outcome for lettuce is not surprising because, like brassica species in general, lettuce is known to be a heavy metal trace element accumulator. So, rather than despair and give up gardening, the Orto Insorto collective not only persisted in finding ways to produce vegetables (regrettably using phytoremediation, which is generally ineffective, as pointed out in Chapter 6) but took the opportunity as well to denounce openly the pollution that capitalism brings. They did so by putting up a large banner facing the nearby road (Figure 7.4).

The collective of the Forte Prenestino garden, safely ensconced within a well-established squat, has other uses that mainly address information and skill sharing, whether or not in terms of self-education or pedagogical activities for children in local schools. The suite of vegetables grown mostly coincides with that of Orto Insorto, save for some more specialised crops like hemp and stevia (a traditional Guaraní cultivar, from whose leaves a sweetener is extracted).

Research conducted in 2014 focused on lead as well as arsenic (Engel-Di Mauro 2018). The soil lead levels were on the lower end of the 2011 soil test results at

FIGURE 7.4 Anti-capitalist banner placed by local activists on the fence surrounding Orto Insorto, Rome (photo by Salvatore Engel-Di Mauro 2014; the banner reads: "1160 ppm Lead; Capitalism Kills")

Orto Insorto (81–523 ppm), but still beyond safe levels. The total soil arsenic content (17–32 ppm) also exceeded the maximum allowable levels (14 ppm according to the US EPA). No lead or arsenic contamination was shown in gardening inputs, but the levels in rainwater and dust were worrisome. In the case of lead, there was virtually no transfer to the vegetables, indicating that soil lead and lead-laden dust simply could not get into the vegetables. In the case of soil lead, this was expected because of soil pH converging on neutral at all sites. The fate of the lead deposited from the air remains a puzzle requiring further funding at least to include a much larger sample size (perhaps it is an issue of airborne particle size being associated with specific elements). In the case of arsenic, there were issues of contamination in the vegetable leaves (16–41 ppm) and, except for vegetables grown in imported soil, in tomatoes and string beans (14–20 ppm), well beyond the safety limits set by the US Centers for Disease Control and Prevention (0.005 ppm) and much higher than can be derived from root absorption (under given soil conditions). The study involved only four sites in two gardens, so the results remain preliminary. Nevertheless, there is, given prior results, a risk of exposure by inhalation and ingestion of soil particles as well as dust from outside the garden. The possibility of airborne contamination for vegetables, evident at least in the case of arsenic, should also be investigated with more extensive research.

The cases of Forte Prenestino and Orto Insorto are better appreciated in a wider context, relative to historical changes as well. While specifically communal gardening (with explicit political goals) is relatively recent in Rome and still somewhat rare, it is an important contrast to the ways urban gardening has been promoted historically. Between the great wars, it was introduced as an obligatory after-work activity by the Fascist regime. Since World War II, producing food in small urban allotments has been largely confined to poverty-stricken elderly migrants from Southern Italy. At most, such gardens help reduce some expenditure on vegetables and/or fulfil pastime needs for pensioners. In any case, official institutions have virtually ignored city-grown food until recently (Cioni 2012; Tei et al. 2010).

Wider social arrangements have also been changing, making for political openings as well as challenges to urban food production. Over the last two decades, the metropolitan area has witnessed demographic rebound and growth (above national trends) accompanying a rise in houselessness and uninhabited buildings. There was a 12.4 per cent expansion in urban area between 2000 and 2010, traceable to real estate speculation and evident as a displacement effect of rising rents. Interest in and spread of community gardens have emerged in this expansionary yet deprivation-inducing context (Mudu and Marini 2018). Community garden initiatives have been partly spurred by public sensitivity to environmental issues (including a rise in localism, see Chapter 5) and dissatisfaction with conventional food, plagued as it is by contamination and migrant-exploitation scandals. To some extent, community gardening is also a response to rising penury (Formisani 2011; Pinto, Pasqualotto, and Levidow 2010) as well as an outcome of struggles over the control of urban space, culminating in the establishment of self-managed spaces (Mudu and Marini 2018). The growth in population and community gardens can also be traced to Rome's unique structure. This has involved, among other factors, much social diversity even within the same multiple housing units and pervasive informal systems of economic activity (Agnew 1995; Mudu 2006). These factors are likely behind the tendency for urban community gardens in Rome to be set up without state incentives or support and for their location to cover most urban socio-spatial zones (Pacione 1998; Mudu and Marini 2018).

In such an overall context, community gardens are entirely novel because they are grassroots initiatives designed to promote diffuse control over the city and from below. Often women-managed, they are mainly formed by precarious and unemployed workers, poorer pensioners, and migrants, with substantial radical squatter involvement (Attili 2013; Bartoletti 2014; Cioni 2012; Del Monte and Sachsé 2018; Mudu 2006; Mudu and Marini 2018). They can, especially in the case of squats, represent instances of practicable alternatives (or prefiguration) to capitalist relations (Ledant 2017; Tornaghi and Dehaene 2020), with repercussions such as the decommodification of land and re-orientation of local economies towards fulfilment of people's needs. Municipalities, in response to this upsurge, are beginning to introduce legal frameworks that go little beyond enabling and regulating land access. Much press fanfare and plenty of conferences are emerging that seem intent on co-opting urban gardening to business ends or on steering it into another source of free labour to compensate for cuts in social services (Attili 2013). Even if garden allotment regulation, finally introduced in 2015 (Coletti 2016), were to facilitate the establishment of community gardens, urban gardening projects face the difficulty of substantive regulatory divergence among 15 different administrations and of shifting policies from a mayor's office rocked by scandals and characterised by frequent personnel change.

Community gardens may be forced into being short-lived expressions of egalitarianism, may not be providing more than marginal increases in food access, and may also imply additional health risks. However, they are of great importance in forging ways to involve people in the direct and participatory management of urban spaces, independently of and in opposition to businesses and their variously legalistic incarnations in the local state. Just as importantly, as the Orto Insorto collective cleverly showed, community gardens can form outlets for the denunciation of capitalist policies that put people's health at risk and, if sufficient resources and people are involved, for the development of environmental monitoring from below.

Reinventing urban farming in Havana

To recapitulate the case studies so far, Tamale and Dar es Salaam feature long-standing and widely practised food production spaces, while Rome, where prior gardening was largely private, is witnessing new, sometimes collective and grassroots-initiated forms of urban food production that call into question mainstream city politics. Institutional support is at most feeble if not absent in all those situations. This could not contrast more with the relatively unique case of

Havana, where the first urban farm (*organopónicos*) was set up for civilian uses in 1991 (Koont 2011, 25). Food production long existed in Havana but was institutionally suppressed and relegated to backyards after the 1959 Revolution, until a late 1980s revival. Because of US-imposed embargoes, a high proportion of Havana's land use has been devoted to growing food, about 40 per cent (Ramamurthy and Kazi 2014). Most of it is managed by the state in co-operative small farms. However, since the early 1990s, these have been slowly decentralised into smaller neighbourhood gardens managed by their cultivators—*grupos de parceleros* (Fernandez 2017).

The city, currently inhabited by about 2.1 million people over roughly 782 km^2, is a major port and commercial centre over karst topography and traversed by 11 short, low-flowing streams and the 50-km-long Rio Almendares (Febles-González et al. 2012), which has been contaminated with heavy metals (e.g. cobalt, chromium, lead) by an upstream smelter and landfill (Olivares-Rieumont et al. 2005). Like Tamale, the climate is tropical savanna with marked seasonality. However, the main meteorological influence (and hazard) comes from droughts and hurricanes, ever more frequent and intense, as well as increasing sea levels. The state-led introduction and spread of agroecological techniques, however, has not only been transforming urban ecosystems by raising agrodiversity, reducing exposed soil surfaces, and decreasing agrochemical contamination, but it also has resulted in improved resilience of food-producing areas to the onslaught of more frequent and extreme weather events related to global warming (Altieri and Funes-Monzone 2012). High-intensity rainfall can still cause much flooding, due to insufficient draining and channelling capacity. Groundwater and aquifers are also increasingly affected by salt-water intrusion resulting from sea-level rise, as in the case of Dar es Salaam. Industrial plants and other installations amount to 197 point-sources of pollution and diffuse sources mostly associated with motorised vehicles, though cases of smog tend to be rare and particulate matter levels, in great contrast to Rome, are usually low (Placeres, Melián, and Toste 2011). Lead and other trace element contamination is a concern downwind of industrial point sources, while contamination via vehicular traffic seems understudied (Álvarez et al. 2017).

Most soils are red and chalk brown earths, moderately alkaline and, typically, with nutrient levels sufficient for fruit trees, pastures, and cane sugar. Towards the coastal areas, there are cases of exposed karst (Placeres, Melián, and Toste 2011). In some cases, soils have been contaminated through open solid waste dumping over the past century. Some areas where cultivation is being considered are therefore affected by contamination legacies (Rizo et al. 2012). Lead and zinc are particularly high in industrial zones, and cobalt and nickel are in part geogenic. Aside from industrial areas, there are school grounds and city parks with high levels of cobalt, nickel, zinc, and copper from human sources, which may be due to aerial deposition from industrial and vehicular emissions (Rizo et al. 2011). Given the prevailing alkaline conditions of soils and the widespread use of compost in urban cultivation, the main contamination threats are likely from airborne sources and possibly through watering and storm-related runoff.

Urban cultivators have had to face these and other biophysical challenges and will be facing greater difficulties with the effects of climate change. Among the more salient ones is the scarcity of water appropriate for crop production (Koont 2011, 180). Salt-water intrusion into aquifers and the contamination of the Rio Almendares exacerbate this problem. Urban food producers, on the other hand, have been bringing about major ecological changes. One is the foregoing of agro-chemicals and farming machinery in favour of organic farming techniques and agroecological applications. Such changes include overhauls in land-use and land-access distribution (Fernandez 2017). Another is contributing to urban reforestation (including fruit trees) and green space expansion more broadly, as urban food production units also participate in a national greening programme (Koont 2011, 175–6). Urban heat island effects and pollutant dispersal can be radically reduced this way, while moisture can be retained more effectively, and flooding magnitude could also be mitigated. Agrodiversity, if not total biodiversity, may also be increasing as a result of the spread of ecologically sustainable farming.

The blossoming of urban farming by the 1990s is the fruit of heavy state promotion combined with initiatives from below (Chaplowe 1998; French, Becker, and Lindsay 2010; Marshalek 2017). The virtual disappearance of small, private gardening and animal husbandry by the 1960s gave way to greater dependence on rural farming. By the late 1980s, under duress from a continuous US embargo, military threats, and attempts on Castro's life, a sea-change occurred. The disappearance of the USSR resulted in the loss of crucial sources of raw materials and machinery that threatened, among other things, the ability of food provisioning and the state's capacity to ensure food access. Drastic measures (such as food rationing and conversions to low-input farming) were put in place that, after much resistance from within and outside the Communist Party and the state, produced much compromise among various factions in how food is to be produced and distributed, accompanied by some decentralisation in decision-making processes (Hearn and Alfonso 2012; Premat 2012). Grassroots pressures were equivalently important in this major change of course. The government was also, to some extent, compelled to concentrate on urban food production because about 80 per cent of Cubans live in cities. The now celebrated, larger, and more commercially oriented urban farms have been supplemented by smaller usufruct-based *parcelas* on public land and home *patios* (now enjoying official appreciation) that provide subsistence as well as supplement private earnings (Altieri et al. 1999; Koont 2011, 165; Levins 2005; Machado 2017; Rosset and Benjamin 1994).

Urban food production in Havana and in Cuba generally is exceptional. Not only has it been reintroduced and supported by state institutions, but it is also integrated into wider agricultural planning, including peri-urban areas. This level of coordination is possible when the national state retains tenure over most land, private enterprise is restricted, and profitability is subordinated to a primary directive of feeding people. Furthermore, the policies of the Cuban government over previous decades have been crucial in establishing research and extension structures, higher general educational levels, skilling processes important in confronting the sudden economic downturn of the 1990s, and development and implementation of technical innovations and improvements for urban farming. It was also state institutions that played key roles in providing the inputs, material incentives, and moral inducements (e.g. patriotism rhetoric) to diffuse agroecological and organic farming methods (Fernandez 2017; Koont 2011, 8; Premat 2012).

Estimates on the number of urban cultivators are contradictory. Some claim that 50,000 and others that 90,000 people are involved in urban food production. Regardless, this remains a large number of people who have become involved. It is especially noteworthy that 7 per cent of Cuba's workers are formally employed in urban agriculture (Koont 2011, 191). Some of the reasons lie in economic benefits potentially gained by cultivators. Roughly 40 per cent of Havana's food production goes to marketing surplus (González Novo and Castellanos Quintero 2014). The result of this combination of inducements is that vegetable and fruit production has reached levels hovering at or exceeding minimum levels to cover the city's nutritional needs. Much of what is grown, especially on the *organopónicos* and intensive gardens, covers everyday popular culinary needs and abides by agroecological principles, diffused also via state extension programmes and farmers' own innovations (Leitgeb et al. 2011). Beans, gourds, lettuce, melons, plantains, tomatoes, and watermelons, for instance, are grown using composts and more concentrated organic fertilisers and in combination with herbs and other plants helpful in warding off pests (French, Becker, and Lindsay 2010, 158; Leitgeb, Schneider, and Vogl 2016).

Under current circumstances (international as well as national), urban cultivation can thereby complement the production of staple crops in the countryside, where organic techniques have been diffused institutionally. This way, shortfalls in food production and access can be overcome more easily, at the same time that there emerges and develops greater self-reliance in society (part of building social capital) as well as more ecologically sensible practices. The fate of urban cultivation is said by some to hinge on prevailing economic trajectory, which responds to modifications of US policies and to wider, global capitalist shifts. The recent rapid decline in larger gardens (including *organopónicos*) was in part to give way to more lucrative tourism and manufacturing land use (French, Becker, and Lindsay 2010, 159). This appears to indicate an inverse relationship between economic hardship and urban food production, in ways that resemble developments in many global North cities. However, the continuous rise in the number of smaller *parcelas* and *patios* (Koont 2009) contravenes any notion of overall urban food production decline. Instead, what is happening is that state policies and movements from below support the spread of more decentralised practices as well as the displacement of higher revenue-generating activities to peri-urban areas, even as they aid economically in surviving relentless US embargo pressures.

Productivity may be relatively high, especially when compared with gardening in cities in the global North, but prospects may be tied to what people decide to do as the harshness of the 1990s gives way to improved living conditions and possibly greater availability of fossil fuels and agrochemicals (Koont 2011), though the 2020 conjuncture of even harsher US government sanctions under pandemic conditions may serve as a brake on this possible tendency. Nevertheless, it is important to note that urban food production and countryside farming have been generally insufficient to ensure food availability nationally. Consequently, Cuba continues to depend highly on imported foods (up to 40 per cent of total food consumed, if not more). Much is being made of this difficulty as part of an effort to show the alleged failures of the socialist state in Cuba (Machado 2017). The actual failure lies in critics' penchant for ignoring the overall situation. It can only be surprising to tunnel vision enthusiasts that an island country hit by a decades-long embargo is unable to exit its import dependence and general colonial plantation legacies, which have been reinforced through decades of trade with the USSR and allied states. Instead, Cubans are supposed to achieve what virtually no other country in the Caribbean (and beyond) has yet to achieve (Levins 2004). But the farcical nature of such critique is made even more evident when comparisons are made with the most capital-rich countries. For example, another island nation, the UK, imports 50 per cent of its food (DEFRA 2017). Coupled with more than a million people suffering from hunger in the UK, one could be forgiven for saying, and with much greater confidence, what an utter failure liberal democracy is.

The overlooked difficulty with fulfilling food demands in Cuba is that much of the imported food is composed of crops that cannot be adequately grown under tropical island conditions, such as cereals and soybeans, that is, unless the state were to resort to fossil fuel-derived agrochemicals Cuban farming is now renowned for avoiding. Dietary patterns are also tough to change when the prerogative is ensuring all Cubans' nutritional needs are met. This is a prerogative that, for instance, does not exist in the US, where hunger persists amidst one of the highest and most diverse food production levels in the world. One could also argue that an overwhelmingly urban society needs to undergo major change over a couple of generations to transform itself into an agrarian society that uses fossil fuels sparingly. What should be taken as much more consequential is that the benefits of urban food production have not been evenly spread in Cuba. Social justice issues persist due to racialised class disparities that hinder the attainment of equal food access for Afro-Cubans (Lowell and Law 2017, 112–13).

There are still other, and to us more important, international and contextual aspects that must be considered. In part, the fate of urban food production and the ecosystem impacts it implies in Habana and elsewhere in Cuba is tied to newer linkages being forged since the early 1990s. One of them is with the People's Republic of China (PRC). Its capitalist re-orientation notwithstanding, the PRC has played an important role in the transfer of technical know-how and materials helpful towards the development of low-input, organic farming methods (INIFAT 2010, 11, 18). Moreover, there are cultural dynamics within Cuban society that have in some ways facilitated the urban cultivation renaissance, aside from internationally derived pressures. Some of the early and persisting forms of highly intensive urban cultivation over inhospitable, tiny spaces are traceable to the gardening practices within Cuba's long-standing Chinese diasporic community (see also Koont 2011,

182). Perhaps, then, it is not surprising that General Sio Wong, 1959 revolution veteran of Sino-Cuban background and Party leader, has been a main proponent of urban food production and agroecological principles (Koont 2011, 25–6; Premat 2012, 11). One urban cultivation method, promoted by institutional experts and the main basis of *organopónicos*, is to mix sediment and composted materials to help establish growing areas over virtually uncultivable city surfaces. These are also part of traditional expertise and everyday gardening practices, but they appear not to be recognised in institutional publications. We will see below that similar techniques are not uncommon among urban cultivators in Chongqing, where food production in the city receives no government support except largely as a cultural prop.

Producing food in the interstices of Chongqing's building boom

Arguably, much larger, recently industrialising cities may be more representative of trends, especially in the global South. To exemplify this, at least to some extent, we turn to Chongqing, a city that, like its country, the PRC, has become the paragon of (surreal) excesses and has been much maligned by all sorts of figures, including on the left. Chongqing, in some ways, encapsulates the insanity of most current urban trends (with no apologies to transition towns) and their intimate linkages to global capitalist relations, which continue to posit profit over life and to diffuse and impose capitalist decision-making processes over the heads of majorities. To tell an ecosocial story of Chongqing, we must rely, as in the above cases, on fragments cobbled from disparate sources and deduced from generally known processes, aside from the results of investigations involving one of us.

Lodged at the confluence of the mighty Yangzi (Chang Jiang) and Jialing rivers and partly surrounded by farmland, the city of Chongqing is built on and around hills and floodplains, its skyline rising and ebbing accordingly. The climate, monsoonal and humid subtropical, affords mild to very high temperatures (8°C to 28°C) and copious amounts of rainfall (1,108 mm per year). Frequent fog related to temperature inversions means that airborne pollutants remain trapped at lower elevation (much as happens in places like Los Angeles, magnifying air pollutant concentration. The landscape would be lush with vegetation were it not held back by a frenzied pace of construction since 1997. At that point, Chongqing, a city of about 6 million to 8 million inhabitants—perhaps even 10 million, counting transient people or those without 户口簿 (*hùkǒu bù*), household registration, and associated residence permit—became a municipality answering directly to the national government. This made surrounding farmland officially urban (under municipal control), as the administrative status changed. The legal consequence is farmland under municipal ownership, as opposed to local "collective" ownership (village- or town-level governing institutions). It can therefore be leased to private interests, and farmers may lose their land rights (Brown and van Nieuwenhuizen 2016; Smith 2014). The administrative move—with all its destructive paradoxes—is part of efforts to raise the economic profile of China's inland western provinces. Urban expansion has been faster there than the national average, despite comparatively modest population growth, even accounting for migration from rural areas. The main result has been the rapid growth of industrial production, including coal-fired plants to enable such activity (Rock et al. 2017).

This, to a large extent, explains how an area of 81,000 km^2, where about 31 million people live, has suddenly been opened up directly to private enterprises, making farmland conversion (rather, destruction) lucrative for both interested capitalists and local state officials (Smith 2014), in spite of legal protections for farmland introduced in 1994 (Lang and Miao 2013, 8). Yet, rurality (especially rural migrants) and peasants (农民, *nóngmin*) are generally associated with backwardness, culturally fanning the flames of mining, manufacturing, and urban encroachment, viewed often as part of progress. Farming, becoming urban by default since 1997, is out of place, a practice from which many prefer being dissociated. Under such conditions, producing food in Chongqing can be understood as an act much against the grain, as it were, immediately political and counter-cultural even if not intended as such and based on fulfilling subsistence needs. Unsurprisingly, many cities, historically or increasingly dependent on foreign investment and export-oriented production (like Chongqing), succumb to similar pressures and, where administrative frameworks converge, exhibit rapid farmland disappearance. Where urban food production has been officially promoted in Beijing, Shanghai, and Fangshan, actual productivity is insignificant, even as efforts concentrate on creating urban agribusinesses—the emphasis is largely educational or cosmetic, to attract tourists (Lang and Miao 2013).

The environmental repercussions of recent changes in Chongqing have been rather dire. Particulate matter (PM_{10}) spiked by the early 2000s (150 μm^{-3}), but mean annual values eventually diminished by 2012 to still relatively high levels (90 μm^{-3}). The finer dust fraction ($PM_{2.5}$) reaches similar averages (ca. 75 μm^{-3}) and is largely from coal combustion. Trace elements, such as nickel and chromium, are embedded in some of these particles and originate from industrial emissions (Chen et al. 2017). The increase in vehicular traffic and massive roadway construction have raised contaminant loads even further, particularly cadmium, iron, and lead, in both air and, by way of runoff, water. Both urban soils and local waterways are consequently enriched in such contaminants (Wang et al. 2013). In addition, local coal plant and other emissions have given rise to acid rain that, at one point, reached values as low as pH 4.5 (Gao et al. 2001). If such tendencies persist and contribute to acidifying soils, heavy metals could become more soluble and get into plant life, including vegetable crops. Surrounding farmland has been similarly impacted by contamination due to industrial expansion and mining, affecting local produce with elevated cadmium and lead levels (Yang, Li, and Long 2007).

Accompanying such rapid urban ecosystem transformation, land commodification and speculation pressures have intensified, curtailing land access for most. Rural (and internal urban) displacement and rising poverty levels (especially among pensioners) have created a situation where many seek to cultivate whatever plots of land they can access so as to secure food for their households (Brown and van Nieuwenhuizen 2016, 89–90; Cao 2014; Lai 2002; Liu et al. 2016). Similar findings are reported for Wuhan (Horowitz and Liu 2017, 211–12), where food producers are mostly women, people older than 30 years, and low income. In Chongqing, many have little farming experience and have learned quickly from others with farming backgrounds. What is grown overcomes potential food shortages and/or enables savings on groceries to be applied to other household expenses such as on healthcare, rent, and the like (Rock et al. 2017).

Unlike in cities similar to Wuhan (Horowitz and Liu 2017), cultivation seems not very common on rooftops and balconies, as much as over courtyards and in large parcels. The latter is due to the open spaces temporarily created through the rough and tumble of land speculators. Urban cultivators typically scope out lots where construction has been halted or wherever there are spaces available, including city parks and riverbanks. They squat such land and come to informal or tacit agreements with government officials, owners, or building managers. The land is then subdivided into individual plots among neighbours in a sort of allotment process from below (the dynamics of which remain to be studied). With such arrangements, food production conditions are inherently tenuous. Many cultivate on plots rarely larger than 25 m^2 on marginal spaces where construction is in the planning phase or has been interrupted (Figure 7.5).

FIGURE 7.5 An area used by gardeners where construction has been halted; Chongqing, Jiang Bei District (photo by Salvatore Engel-Di Mauro 2015)

Areas under construction tend to be large expanses, tens of hectares, of partly excavated and exposed and partly re-vegetated land. Reoccupation by pioneer herbaceous plants, aided by favourable climate, seems not to take very long. Cultivators may need to clear plots prior to planting with vegetables. Those who occupy such temporarily abandoned projects cultivate plots on average for little more than 2 years before they are forced to leave. This is too short a time span for adapting to local environmental conditions, engaging in land-use planning, and dedicating effort to improve cultivation prospects. Often, the land accessed is steeply sloped and more susceptible to erosion, especially in cases where topsoil has been removed to make room for future building and street foundations. Where soils are not truncated, they may be thin as a result of their pre-existing topographical situation on slopes, which makes such soils prone to erosion even without human intervention. This also means they tend to be less fertile and to have less water-holding capacity. Other kinds of local soils may be silty or clayey and iron-rich, which also tend to be low in fertility and potentially low in permeability (contributing to heightened floods). Most of the soils investigated by Engel-Di Mauro (see Rock et al. 2017) were sandy and much richer in organic matter, but this was partly because of sediment and organic waste additions by cultivators.

Subsistence-oriented production prevails, and inputs tend to be free of agrochemicals. Local piped water supplies, which may be contaminated by organic and inorganic pollutants as well as pathogens, are used for manual irrigation by bringing water in containers to the vegetables. Some use night soil, but, for the most part, there is reliance on a variety of organic composts, ranging from household consumables or production residues (e.g. leftovers and soy-milk pulp) to varieties of manures from raised fowl (chickens, geese, and ducks) and composted weeds. Crop diversity ranged from as little as one to as many as 23 different crops in a single plot. Among the most popular vegetables were long beans, cucumber, maize, courgettes, chillies, and water spinach. Many cultivators use polycultural cropping techniques. Save for flat areas unaffected by construction, soils tended to be less than 20 cm deep, which severely limits rooting. Average colour was sufficiently dark in longer-cultivated plots to suggest moderate levels of soil organic matter development. Otherwise, soils tended to be sandy with blocky structures and loose to friable consistence, which makes for relatively high permeability and, where soil is thin and on underlying concrete, susceptibility to erosion. In construction areas, soils were truncated (i.e. the topsoil was largely removed), which implies low nutrient and soil organic matter levels.

With no institutional support, food production in Chongqing occurs largely within exposed excavated areas, public forested parks, or in cramped blocks between buildings (Figure 7.6), where sediment has been mixed with organic wastes to form thin soil. It may be on foundations of razed structures or in areas being prepared for new construction. In this latter case, land access, already provisional because of lack of tenure or any state guarantees, is fleeting, and cropped areas can be destroyed at any moment. The situation may represent an ephemeral process linked to a conjuncture of rural displacement, changing central-government policies, and

FIGURE 7.6 A gardening area set up by nearby residents amidst a block of flats, metropolitan railway, and elevated motorway; Chongqing, Yu Zhong District (photo by Salvatore Engel-Di Mauro 2015)

rapid urban expansion and intensification. Nevertheless, the food growing reflects remarkable neighbourhood-based organising and self-management among essentially landless urban dwellers (only one land owner was noted in the study), replete with informal self-education and knowledge-sharing processes.

Biophysically, the result is a re-vegetation of otherwise bare and therefore more erodible surfaces, contributing to reducing runoff and airborne dust. Within short time spans, not only is topsoil being reconstituted and soil organic matter developed (raising fertility levels), but soils are being stabilised by terracing on steep slopes. Some are even being manufactured over concrete within very short time spans, by recycling organic wastes and otherwise loose sediment. Habitats are simultaneously being created for other organisms so that cultivated areas may raise species richness (though not necessarily restoring former species). The extent of these biophysical effects may be temporary and too geographically delimited, but they point to what could be done to reverse currently destructive processes, and done so in ways that could foster local empowerment and participatory self-management. A possible exception to these positive developments is the establishment of cropping areas in city parks dominated by woods. The felling of trees to clear areas for cultivation will likely reduce potentials for GHG storage and moisture retention, among other net negative outcomes. Another potential problem to be explored is exposure to and/or net containment of contaminants, especially from air pollution generally but also through abandoned factories or processing plants, construction zones, and proximity to major motorways.

The current food–city relationship in Chongqing seems to represent an ephemeral assemblage of haphazardly arranged vegetable cultivation areas that emerged as a result of converging capitalist and Communist Party directives in creating economic hubs in the interior of the country so as to expand internal command as well as to promote business and government interests abroad (a combination of capital accumulation and political power consolidation). Biophysically, a situation has developed in which sometimes risky practices (e.g. the use of night soil) combine with increasing levels of environmental pollutants that may expose (especially women) cultivators to elevated levels of toxic substances more than otherwise would be the case. The largely itinerant nature of cultivation, resulting from highly inimical politics in land allocation on the side of capitalists, is unlikely to offer the sort of environmental and ecological benefits of durable cultivation areas, such as pollutant mobility abatement and biodiversity enhancement.

Associated social benefits are also likely ephemeral. There may be public health issues derived from consuming contaminated food, aside from concentrated amounts of pollutants, but all such matters need to be verified. The lack of policies to confront, never mind legitimise, urban cultivation is already an obstacle to bringing about clarification. Additional social benefits may nevertheless persist beyond the duration of land access, thanks to the perseverance of cultivators themselves. These come in the form of skill sharing, agronomic education, and lasting relationships forged out of mutual help among different households. As in liberal democracies, the fate of urban cultivation is at the mercy of local business and government as well as cultivators' abilities to resist eviction or to find new cultivation spaces. However, the food produced is of pivotal importance to the viability of the households involved, and this in itself makes of urban cultivation an eloquent statement on the increasing disparities within Chongqing (and beyond) as well as a potential for the discrediting of current state policies through direct political confrontation between cultivators and local authorities.

Local conditions and global prospects

When integrated with other broadly known biophysical processes and findings, diverse expertise-specific studies of the same city can be brought together to illustrate what can be gained by analysing the social as well as the biophysical. This was done here by means of five case studies from Tamale, Dar es Salaam, Rome, Havana, and Chongqing. These urban ecosystems cover several kinds of biophysical contexts and challenges (from tropical to temperate regional climates), differing levels of material well-being and industrialisation, different sorts of institutional policies (ranging from inimical through indifferent to actively supportive), and diverse forms and (ecosocial) impacts of urban cultivation. The point is not to offer an exhaustive analysis of all existing permutations of ecosocial inter-relations but to demonstrate how their mutually transformative natures (or dialectical relationships) in wider biophysical and social contexts can reveal the limits and possibilities of urban cultivation, especially in advancing the case for ecosocialism.

In Tamale, urban cultivation has existed possibly for as long as the town itself and has not been subjected to any debilitating policies. Rather, food production forms the main economic activity that could be ecologically constructive were it not for the agrochemicals and contaminated wastewater employed. The prospects so far are far from ideal. At the same time, social stratification forms the backbone of these ecologically unsustainable practices. In a likely shortened Harmattan season with climate change and mounting pollution problems, existing inequalities and health risks will likely sharpen among the mostly women involved in food production. Local social relations of power will need to be addressed first and foremost if urban food production is to contribute to social and environmental justice, e.g. sharing technical knowledge and crop types, as well as other supports useful in confronting crop-specific productivity challenges due to regional climate change.

Something similar can be said of Dar es Salaam, where urban farming is vast in comparison and highly diversified into large profit-oriented pasture and cropping systems and smaller subsistence-focused vegetable and small-animal operations. There is marked gender differentiation, with men predominantly occupied in commercial production and women mainly involved in subsistence and informal surplus sales and exchanges. There are major biophysical challenges due to a combination of climate change and local pollution of soil and water, deforestation, and paved surface expansion. Some of the contamination and other biophysical challenges have been caused by farming itself, which is largely conventional. Though not fully supported by municipal institutions, urban food production is intimately part of the local economy and is ridden by social stratification. The fate of commercial farming seems contingent on the same sort of logic that propels much of it, which is profitability. As greater pressure is exerted by land speculators and other sorts of land-hungry businesses, commercial farming area may eventually become marginalised. However, given the conditions of most people, whose economic fortunes are typically low, subsistence farming is much more likely to be enduring, especially if food price hikes continue and more urbanites become wage-dependent.

The situation in Rome also points to major social injustice and pollution problems that urban cultivation cannot be counted on to resolve, but for different reasons. Urban cultivation has never been seriously supported institutionally and has never been a main feature of the local economy. In fact, municipal institutions usually get in the way of developing food production in the city. Urban gardening, in our analysis, if carried out as a communal effort to stymie or pre-empt land speculators, is of great immediate political importance in questioning not only persisting social injustice (e.g. in housing), but also in bringing to the fore the ecological and human health destructiveness of the capitalist city. This is also why squatters' movements in Rome are an especially important component of a wider political strategy to overcome capitalist relations.

Havana, in this regard, serves as a sort of beacon of what can be accomplished to improve urban life radically. There are certainly problems with how urban cultivation has been institutionally supported, particularly with respect to the continuing marginalisation of Afro-Cuban communities. It is also far from clear whether or not urban cultivation policies will succeed in promoting communal food production rather than household-level production for private gain. It is also unclear how cultivators will avoid exposure to pre-existing contaminants in their plots, given a lack of capillary technical support in identifying contamination levels and diffusing exposure prevention measures. Nevertheless, Havana shows a way forward unlike any other city examined here in combining social justice with ecological sustainability. The contrast could not be greater with the situation faced by Chongqing, where food production is not only unrecognised institutionally, but also even more precarious than in Rome, contingent upon the whims of construction businesses and their local state and private sector allies. The sheer establishment of urban gardens becomes an implicit political statement about what the city could be like, ecologically and socially.

In sum, where food production contributes to ecological sustainability and is solidly in place or even an intimate part of how a city is culturally conceived (e.g. Tamale), there is no clear attempt at political organising to bring about an end to social inequalities. Where cultivation is highly contingent institutionally and socially, and challenged by multiple and major biophysical challenges, including high levels of pollution (Chongqing, Dar es Salaam, Rome, Tamale), urban food production becomes a fulcrum of antagonism towards capitalist relations, whether explicitly expressed (Rome) or tacitly practised by virtue of circumstances (Chongqing, Dar es Salaam, Tamale). Where urban food production has expanded and is the result of combined institutional and wider social initiatives (Havana), there exists much promise in building an ecosocialist alternative. However, a shifting global conjuncture as well as internal pressures may re-direct the overall project towards less than socialist ends. Putting pressure on the US to end its imperialist designs towards Cuba will be helpful towards supporting the great strides made in Cuba in urban food production and development and application of agroecological principles. Forging international linkages and mutual support associations across such cities, pivoting on the experiences in Havana (and Cuba generally) and any other similar situations, would enable the development of urban cultivation as a global force that contributes to creating a basis for its ecosocialist future. Such action will require coordination among many socio-environmental justice and ecological sustainability projects across national and urban–rural divides.

References

Agnew, J. 1995. *Rome*. Chichester: John Wiley & Sons.

Altieri, M.A., and E.R. Funes-Monzone. 2012. The Paradox of Cuban Agriculture. *Monthly Review* 63(8): 23–33.

Altieri, M.A., N. Companioni, K. Cañizares, C. Murphy, P. Rosset, M. Bourque, and C.I. Nicholls. 1999. The Greening of the 'Barrios': Urban Agriculture for Food Security in Cuba. *Agriculture and Human Values* 16: 131–40.

Álvarez, A., J. Estévez Alvarez, C. Nascimento, I. González, O. Rizo, L. Carzola, R. Ayllón Torres, and J. Pascual. 2017. Lead Isotope Ratios in Lichen Samples Evaluated by ICP-ToF-MS to Assess Possible Atmospheric Pollution Sources in Havana, Cuba. *Environmental Monitoring & Assessment* 189(1): 1–8.

Amin, Samir. 1973. *Neo-Colonialism in West Africa*. Harmondsworth: Penguin.

Angelone, M., T. Corrado, and G. Dowgiallo. 1995. Lead and Cadmium Distribution in Urban Soils and Plants in the City of Rome (Italy). In, *International Conference on the Biogeochemistry of Trace Elements*. Paris: DGAD/SRAE.

Armar-Klemesu, M. 2000. Urban Agriculture and Food Security, Nutrition and Health. In, *Growing Cities, Growing Food: Urban Agriculture on the Policy Agenda. A Reader on Urban Agriculture*, edited by N. Bakker, M. Dubbeling, S. Gündel, U. Sabel-Koschella, and H. de Zeeuw, 99–117. Feldafing: Deutsche Stiftung für Internationale Entwicklung (DSE), Zentralstelle für Ernährung und Landwirtschaft.

Attili, G. 2013. Gli Orti Urbani Come Occasione di Sviluppo di Qualità Ambientale e Sociale. Il Caso di Roma. [Urban gardens as an Opportunity for Environmentally and Socially Improved Development. The Case of Rome] In, *Pratiche di Trasformazione dell'Urbano* [Urban Transformation Practices], edited by E. Scandurra, and G. Attili, 47–67. Roma: Franco Angeli.

Bartoletti, R. 2014. Critical Nature: Regenerating Human Experience and Society Through Gardening. *Sociologia Italiana AIS Journal of Sociology* 3: 9–32.

Bell, S., R. Fox-Kämper, N. Keshavarz, M. Benson, S. Caputo, S. Noori, and A. Voigt, eds. 2016. *Urban Allotment Gardens in Europe*. London: Routledge and Earthscan.

Bellwood-Howard, I., and E. Bogweh Nchanji. 2017. The Marketing of Vegetables in a Northern Ghanaian City: Implications and Trajectories. In, *Global Urban Agriculture: Convergence of Theory and Practice Between North and South*, edited by M.G.A. Winkler-Prins, 79–92. Boston, MA: CABI.

Bellwood-Howard, I., G. Kranjac-Berisavljevic, E. Nchanji, M. Shakya, and R. van Veenhuizen. 2018. Participatory Planning for Food Production at City Scale Experiences from a Stakeholder Dialogue Process in Tamale, Northern Ghana. In, *Integrating Food into Urban Planning*, edited by Y. Cabannes, and C. Marocchino, 292–311. London and Rome: UCL Press and FAO.

Bersaglio, B., and T. Kepe. 2014. Farmers at the Edge: Property Formalisation and Urban Agriculture in Dar es Salaam, Tanzania. *Urban Forum* 25(3): 389–405.

Bisset, J.A., M.M. Rodríguez, Y. Ricardo, H. Ranson, O. Pérez, M. Moya, and A. Vázquez. 2011. Temephos Resistance and Esterase Activity in the Mosquito *Aedes Aegypti* in Havana, Cuba Increased Dramatically Between 2006 and 2008. *Medical & Veterinary Entomology* 25(3): 233–9.

Bogweh Nchanji, E., I. Bellwood-Howard, N. Schareika, T. Chagomoka, J. Schlesinger, D. Axel, and G. Rüdiger. 2017. Assessing the Sustainability of Vegetable Production Practices in Northern Ghana. *International Journal of Agricultural Sustainability* 15(3): 321–37.

Briggs, J., and D. Mwamfupe. 1999. The Changing Nature of the Peri-Urban Zone in Africa: Evidence from Dar-es-Salaam, Tanzania. *Scottish Geographical Journal* 115(4): 269–82.

Brown, K., and S. van Nieuwenhuizen. 2016. *China and the New Maoists*. London: Zed Books.

Brunori, G., and F. Di Iacovo. 2014. Alternative Food Networks as Drivers of a Food Transition. In, *Second Nature Urban Agriculture: Designing Productive Cities*, edited by D.A. Viljoen, and K. Bohn, 244–51. London: Routledge.

Calace, N., L. Caliandro, B.M. Petronio, M. Pietrantonio, M. Pietroletti, and V. Trancalini. 2012. Distribution of Pb, Cu, Ni and Zn in Urban Soils in Rome City (Italy): Effect of Vehicles. *Environmental Chemistry* 9: 69–76.

Cao, H. 2014. *Nouvelles Tendances de l'Urbanisation au Sichuan et à Chongqing: Agglomérations Urbaines et Périmètres Administratifs des Villes* [New Urbanisation Tendencies in Sichuan and Chongqing: Urban Agglomerations and City Administrative Perimeters]. Doctoral dissertation, Paris 7.

Cenci, R.M., D. Dabergami, E. Beccaloni, G. Ziemacki, A. Benedetti, L. Pompili, A.S. Mellina, and M. Bianchi. 2008. *Bioindicatori per Valutare la Qualità dei Suoli di Alcuni Parchi della Città di Roma.* [Bioindicators to Assess Soil Quality in Some Parks of the City of Rome]. Luxembourg: Office for Official Publications of the European Communities.

Certomà, C., and C. Tornaghi. 2015. Political Gardening. Transforming Cities and Political Agency. *Local Environment* 20(10): 1123–31.

Chaplowe, S.G. 1998. Havana's Popular Gardens: Sustainable Prospects for Urban Agriculture. *Environmentalist* 18(1): 47–57.

Chen, Y., S.D. Xie, B. Luo, and C.Z. Zhai. 2017. Particulate Pollution in Urban Chongqing of Southwest China: Historical Trends of Variation, Chemical Characteristics and Source Apportionment. *The Science of the Total Environment* 584–5: 523–34.

Cinti, D., M. Angelone, U. Masi, and C. Cremisini. 2002. Platinum Levels in Natural and Urban Soils from Rome and Latium (Italy): Significance for Pollution by Automobile Catalytic Converter. *The Science of the Total Environment* 293: 47–57.

Cioni, L. 2012. *Orti-Culture. Riflessioni Antropologiche sull'Orticultura Urbana* [Garden-Cultures. Anthropological Reflections on Urban Horticulture]. Thesis, University of Bologna. http://trameurbane.noblogs.org/files/2011/09/orti-culture.pdf (Accessed 10 August 2020).

Coletti, R. 2016. Urban Gardening in Rome. www.tess-transition.eu/urban-gardening-in-rome/ (Accessed 10 August 2020).

Conio, O., and R. Porro. 2004. *L'Arsenico nelle Acque Destinate al Consumo Umano* [Arsenic in Water for Human Consumption]. Genova: FrancoAngeli.

Coulson, A. 2013. *Tanzania: A Political Economy.* Oxford: Oxford University Press.

DEFRA. 2017. *Food Statistics in Your Pocket 2017—Global and UK supply.* London: UK Department for Environment, Food & Rural affairs.

Della Seta, P., and R. Della Seta. 1988. *I Suoli di Roma* [The Soils of Rome]. Roma: Editori Riuniti.

Del Monte, B., and V. Sachsé. 2018. Urban Agriculture: From a Creative Disorder to New Arrangements in Rome. In, *The Urban Garden City. Cities and Nature*, edited by S. Glatron, and L. Granchamp, 271–88. Heidelberg: Springer Cham.

Dongus, S., and I. Nyika. 2000. Vegetable Production on Open Spaces in Dar es Salaam—Spatial Changes from 1992 to 1999. www.cityfarmer.org/daressalaam.html (Accessed 14 July 2020).

Drechsel, P., and S. Dongus. 2010. Dynamics and Sustainability of Urban Agriculture: Examples from Sub-Saharan Africa. *Sustainability Science* 5(1): 69–78.

Engel-Di Mauro, S. 2018. An Exploratory Study of Potential as and Pb Contamination by Atmospheric Deposition in Two Urban Vegetable Gardens in Rome, Italy. *Journal of Soils and Sediments* 18(2): 426–30.

Febles-González, J., M. Vega-Carreño, A. Tolón-Becerra, and X. Lastra-Bravo. 2012. Assessment of Soil Erosion in Karst Regions of Havana, Cuba. *Land Degradation & Development* 23(5): 465–74.

Fernandez, Margarita. 2017. Urban Agriculture in Cuba: 30 Years of Policy and Practice. *Urban Agriculture Magazine* 33: 41–4.

Formisani, E. 2011. Ai Cantieri Verdi Basta una Zappata [A Hoeing Is Enough for the Green Workshops]. *Nuovo Paese Sera* 1(2–3): 16–18.

French, C., M. Becker, and B. Lindsay. 2010. Havana's Changing Urban Agriculture Landscape: A Shift to the Right? *Journal of Agriculture, Food Systems, and Community Development* 1(2): 155–65.

Fusco, D., F. Forastiere, P. Michelozzi, T. Spadea, B. Ostro, M. Arcà, and C.A. Perucci. 2001. Air Pollution and Hospital Admissions for Respiratory Conditions in Rome, Italy. *European Respiratory Journal* 17(6): 1143–50.

Gao, S., K. Sakamoto, D. Zhao, D. Zhang, X. Dong, and S. Hatakeyama. 2001. Studies on Atmospheric Pollution, Acid Rain and Emission Control for Their Precursors in Chongqing, China. *Water, Air & Soil Pollution* 130(1/4): 247–52.

Garcia-Aristizabal, A., E. Bucchignani, E. Palazzi, D. D'Onofrio, P. Gasparini, and W. Marzocchi. 2015. Analysis of Non-Stationary Climate-Related Extreme Events Considering Climate Change Scenarios: An Application for Multi-Hazard Assessment in the Dar es Salaam Region, Tanzania. *Natural Hazards* 75(1): 289–320.

González Novo, M., and A. Castellanos Quintero. 2014. Havana. In, *Growing Greener Cities in Latin America and the Caribbean. An FAO Report on Urban and Peri-Urban Agriculture in the Region*, edited by G. Thomas, 10–19. Rome: FAO.

GsF (Geologia senza Frontiere). 2012. *Geologia dell'Orto Insorto*. Roma: Associazione Geologia Senza Frontiere Onlus.

Gyasi E.A., G. Kranjac-Berisavljevic, M. Fosu, A.M. Mensah, G. Yiran, and I. Fuseini. 2014. Managing Threats and Opportunities of Urbanisation for Urban and Peri-Urban Agriculture in Tamale, Ghana. In, *The Security of Water, Food, Energy and Liveability of Cities*. Water Science and Technology Library, vol. 71, edited by B. Maheshwari, R. Purohit, H. Malano, V. Singh, and P. Amerasinghe, 87–97. Dordrecht: Springer.

Hazelip E. 2014. *Agricoltura Sinergica. Le Origini, l'Esperienza, la Pratica*. [Synergistic Agriculture. Origins Experiences, Practices] Firenze: Terra Nuova Edizioni.

Hearn, A.H., and F.J. Alfonso. 2012. Havana: From Local Experiment to National Reform. *Singapore Journal of Tropical Geography* 33(2): 226–40.

Himberg, L.M. 2016. *Mangroves and Urbanization: Systems in Transition. A Study of Social-Ecological Systems of Mangroves in Dar es Salaam, Tanzania*. Master's thesis, Norwegian University of Life Sciences. https://brage.bibsys.no/xmlui/handle/11250/2398882 (Accessed 12 September 2018).

Horowitz, S.H., and J. Liu. 2017. Urban Agriculture and the Reassembly of the City: Lessons from Wuhan, China. In, *Global Urban Agriculture: Convergence of Theory and Practice Between North and South*, edited by A.M.G.A. WinklerPrins, 207–19. Boston, MA: CABI.

Howorth, C., I. Convery, and P. O'Keefe. 2001. Gardening to Reduce Hazard: Urban Agriculture in Tanzania. *Land Degradation & Development* 12(3): 285–91.

INIFAT (Instituto de Investigaciones Fundamentales en Agricultura Tropical). 2010. *Manual Técnico para Organopónicos, Huertos Intensivos y Organoponía Semiprotegida*. [Technocal Manual for Organopónicos, Intensive Gardens, and Semi-Protected Organopónicos] Septima Edición. Habana: INIFAT.

Justin, M.G., J.M. Bergen, M.S. Emmanuel, and K.G. Roderick. 2018. Mapping the Gap of Water and Erosion Control Measures in the Rapidly Urbanizing Mbezi River Catchment of Dar es Salaam. *Water* 10(1): 1–N.

Karg, H., P. Drechsel, E.K. Akoto-Danso, R. Glaser, G. Nyarko, and A. Buerkert. 2016. Foodsheds and City Region Food Systems in Two West African Cities. *Sustainability* 8: 1175.

Kiunsi, R. 2013. The Constraints on Climate Change Adaptation in a City with a Large Development Deficit: The Case of Dar es Salaam. *Environment & Urbanization* 25(2): 321–37.

Koont, S. 2009. The Urban Agriculture of Havana. *Monthly Review* 60(8): 44–63.

Koont, S. 2011. *Sustainable Urban Agriculture in Cuba*. Gainesville: University Press of Florida.

Lai, H.H. 2002. China's Western Development Program: Its Rationale, Implementation, and Prospects. *Modern China* 28(4): 432–66.

Lang, G., and B. Miao. 2013. Food Security for China's Cities. *International Planning Studies* 18(1): 5–20.

Langan, M. 2017. *Neo-Colonialism and the Poverty of 'Development' in Africa*. New York: Palgrave Macmillan.

Ledant, C. 2017. Urban Agroecology in Rome. *Urban Agriculture Magazine* 33: 31–3.

Lee-Smith, D. 2010. Cities Feeding People: An Update on Urban Agriculture in Equatorial Africa. *Environment and Urbanization* 22(2): 483–99.

Leitgeb, F., F.R. Funes-Monzote, S. Kummer, and C.R. Vogl. 2011. Contribution of Farmers' Experiments and Innovations to Cuba's Agricultural Innovation System. *Renewable Agriculture and Food Systems* 26(4): 354–67.

Leitgeb, F., S. Schneider, and C.R. Vogl. 2016. Increasing Food Sovereignty with Urban Agriculture in Cuba. *Agriculture and Human Values* 33: 415–26.

Levins, R. 2004. Cuba's Biological Weapons. *Capitalism Nature Socialism* 15(2): 31–3.

Levins, R. 2005. How Cuba Is Going Ecological. *Capitalism Nature Socialism* 16(3): 7–25.

Liu, W., M. Dunford, Z. Song, and M. Chen. 2016. Urban—Rural Integration Drives Regional Economic Growth in Chongqing, Western China. *Area Development and Policy* 1(1): 132–54.

Loram, A., K. Thompson, K. Warren, and K.J. Gaston. 2008. Urban Domestic Gardens (XII): The Richness and Composition of the Flora in Five UK Cities. *Journal of Vegetation Science* 19: 321–30.

Lowell, J.T., and S. Law. 2017. Sustainability's Incomplete Circles: Towards a Just Food Politics in Austin, Texas and Havana, Cuba. In, *Global Urban Agriculture: Convergence of Theory and Practice between North and South*, edited by A.M.G.A. WinklerPrins, 106–17. Boston, MA: CABI.

Machado, M.R. 2017. Alternative to What? Agroecology, Food Sovereignty, and Cuba's Agricultural Revolution. *Human Geography* 10(3): 7–21.

Mahugija, J.A., P.E. Chibura, and E.H. Lugwisha. 2018. Residues of Pesticides and Metabolites in Chicken Kidney, Liver and Muscle Samples from Poultry Farms in Dar es Salaam and Pwani, Tanzania. *Chemosphere* 193: 869–74.

Mahugija, J.M., A. Kayombo, and R. Peter. 2017. Pesticide Residues in Raw and Processed Maize Grains and Flour from Selected Areas in Dar es Salaam and Ruvuma, Tanzania. *Chemospher* 185: 137–44.

Mahugija, J.M., F.A. Khamis, and E.J. Lugwisha. 2017. Assessment of Pesticide Residues in Tomatoes and Watermelons (Fruits) from Markets in Dar es Salaam, Tanzania. *Journal of Applied Sciences & Environmental Management* 21(3): 497–501.

Marshalek, F. 2017. Cuban and Danish Agriculture, the Rochdale Principles, and the Renovation of Socialism. *Human Geography* 10(3): 22–40.

Mbuligwe, S.E., and G.R. Kassenga. 1997. Automobile Air Pollution in Dar es Salaam City, Tanzania. *The Science of the Total Environment* 199(3): 227–35.

McCrickard, L.S., A.E. Massay, R. Narra, J. Mghamba, A.A. Mohamed, R.S. Kishimba, L.J. Urio, N. Rusibayamila, G. Magembe, M. Bakari, J.J. Gibson, R.B. Eidex, and R.E.

Quick. 2017. Cholera Mortality During Urban Epidemic, Dar es Salaam, Tanzania, August 16, 2015-January 16, 2016. *Emerging Infectious Diseases* 23: S154–7.

Mkalawa, C. 2016. Analyzing Dar es Salaam Urban Change and Its Spatial Pattern. *International Journal of Urban Planning and Transportation* 31(1): 1138–50.

Mkwela, H.S. 2013. Urban Agriculture in Dar es Salaam: A Dream or Reality? *WIT Transactions on Ecology and The Environment* 173: 161–71.

Mlozi, M. 1996. Urban Agriculture in Dar es Salaam: Its Contribution to Solving the Economic Crisis and the Damage It Does to the Environment. *Development Southern Africa* 13(1): 47–65.

Mlozi, M. 1997. Impacts of Urban Agriculture in Dar es Salaam, Tanzania. *The Environmentalist* 17(2): 115–24.

Moses, G. 2015. *The Impacts of Community Gardens on Household Food Security: The Case of Chikato Ward 8, in Tongogara District*. Gweru: Geography and Environmental Studies, Midlands State University.

Mougeot, L.J.A. 2015. Urban Agriculture in Cities in the Global South: Four Logics of Integration. In, *Food and the City, Histories of Culture and Cultivation*, edited by D. Imbert, 163–93. Washington, DC: Dumbarton Oaks.

Mtoni, Y., I. Mjemah, C. Bakundukize, M. Camp, K. Martens, and K. Walraevens. 2013. Saltwater Intrusion and Nitrate Pollution in the Coastal Aquifer of Dar es Salaam, Tanzania. *Environmental Earth Sciences* 70(3): 1091–111.

Mudu, P. 2006. Patterns of Segregation in Contemporary Rome. *Urban Geography* 27(5): 422–40.

Mudu, P., and A. Marini. 2018. Radical Urban Horticulture for Food Autonomy: Beyond the Community Gardens Experience. *Antipode* 50(2): 549–73.

Mwalukasa, M. 2000. Institutional Aspects of Urban Agriculture in the City of Dar es Salaam. In, *Growing Cities, Growing Food: Urban Agriculture on the Policy Agenda. A Reader on Urban Agriculture*, edited by N. Bakker, M. Dubbeling, S. Gündel, U. Sabel-Koschella, and H. de Zeeuw, 147–59. Feldafing: Deutsche Stiftung für Internationale Entwicklung (DSE), Zentralstelle für Ernährung und Landwirtschaft.

Mwegoha, W.J.S., and C. Kihampa. 2010. Heavy Metal Contamination in Agricultural Soils and Water in Dar es Salaam City, Tanzania. *African Journal of Environmental Science and Technology* 4(11): 763–9.

Ndetto, E., and A. Matzarakis. 2013. Basic Analysis of Climate and Urban Bioclimate of Dar es Salaam, Tanzania. *Theoretical & Applied Climatology* 114(1/2): 213–26.

Nkrumah, K. 1965. *Neo-Colonialism: The Last Stage of Imperialism*. London: Thomas Nelson & Sons.

Nugent, R. 2000. The Impact of Urban Agriculture on the Household and Local Economies. In, *Growing Cities, Growing Food: Urban Agriculture on the Policy Agenda. A Reader on Urban Agriculture*, edited by N. Bakker, M. Dubbeling, S. Gündel, U. Sabel-Koschella, and H. de Zeeuw, 67–97. Feldafing: Deutsche Stiftung für Internationale Entwicklung (DSE), Zentralstelle für Ernährung und Landwirtschaft.

Olivares-Rieumont, S., D. de la Rosa, L. Lima, D.W. Graham, K. D-Alessandro, J. Borroto, F. Martínez, and J. Sánchez. 2005. Assessment of Heavy Metal Levels in Almendares River Sediments—Havana City, Cuba. *Water Research* 39(16): 3945–53.

Orsini, F., R. Kahane, R. Nono-Womdin, and G. Giaquinto. 2013. Urban Agriculture in the Developing World: A Review. *Agronomy for Sustainable Development* 33: 695–720.

Pacione, M. 1998. The Social Geography of Rome. *Tijdschrift voor Economische en Sociale Geografie* 89(4): 359–70.

Pinto, B., A. Pasqualotto, and L. Levidow. 2010. Community Supported Urban Agriculture: The Orti Solidali Project in Rome. *Urban Agriculture Magazine* 24: 58–60.

Placeres, M.R., M.G. Melián, and M.A. Toste. 2011. Principales Características de la Salud Ambiental de la Provincia La Habana. [Main Characteristics of Environmental Helath in the Privince of La Havana] *Revista Cubana de Higiene y Epidemiología* 49(3): 384–98.

Premat, A. 2012. *Sowing Change: The Making of Urban Agriculture in Havana*. Nashville: Vanderbilt University Press.

Ramamurthy, K.V., and S.A. Kazi. 2014. Urban Agriculture in Cuba and Exploring Possibilities with Reference to Urban India. *International Journal of Humanities, Arts, Medicine and Sciences* 2: 77–88.

Rizo, O.D., F. Castillo, J. López, and M. Merlo. 2011. Assessment of Heavy Metal Pollution in Urban Soils of Havana City, Cuba. *Bulletin of Environmental Contamination & Toxicology* 87(4): 414–19.

Rizo, O.D., M. Hernández Merlo, F. Echeverría Castillo, and J. Arado López. 2012. Assessment of Metal Pollution in Soils from a Former Havana (Cuba) Solid Waste Open Dump. *Bulletin of Environmental Contamination & Toxicology* 88(2): 182–6.

Rock, M., S. Engel-Di Mauro, S. Chen, M. Iachetta, A. Mabey, K. McGill, and J. Zhao. 2017. Food Production in Chongqing, China: Opportunities and Challenges. *Middle States Geographer* 49: 55–62.

Rosset, P., and M. Benjamin. 1994. Cuba's Nationwide Conversion to Organic Agriculture. *Capitalism Nature Socialism* 5(3): 79–97.

Sappa, G., S. Ergul, F. Ferranti, L.N. Sweya, and G. Luciani. 2015. Effects of Seasonal Change and Seawater Intrusion on Water Quality for Drinking and Irrigation Purposes, in Coastal Aquifers of Dar es Salaam, Tanzania. *Journal of African Earth Sciences* 105: 64–84.

Sattler, M.A., D. Mtasiwa, M. Kiama, Z. Premji, M. Tanner, G.F. Killeen, and C. Lengeler. 2005. Habitat Characterization and Spatial Distribution of *Anopheles sp.* Mosquito Larvae in Dar es Salaam (Tanzania) During an Extended Dry Period. *Malaria Journal* 4: 4. https://malariajournal. biomedcentral.com/track/pdf/10.1186/1475-2875-4-4 (Accessed 12 September 2018).

Sawio, C.J. 1998. Managing Urban Agriculture in Dar es Salaam. https://idl-bnc-idrc.dspacedirect.org/bitstream/handle/10625/23305/108517.pdf?sequence=1 (Accessed 12 September 2018).

Schmidt, S. 2012. Getting the Policy Right: Urban Agriculture in Dar es Salaam, Tanzania. *International Development Planning Review* 34(2): 129–45.

Smith, N.R. 2014. Living on the Edge: Household Registration Reform and Peri-Urban Precarity in China. *Journal of Urban Affairs* 36: 369–83.

Tei, F., P. Benincasa, M. Farneselli, and M. Caprai. 2010. Allotment Gardens for Senior Citizens in Italy: Current status and Technical Proposals. In, *Proceedings of the 2nd International Conference on Landscape and Urban Horticulture*, edited by G. Prosdocimi Giaquinto, and F. Orsini, *Acta Horticulturae* 881: 91–5.

Tibesigwa, B., R. Lokina, F. Kasalirwe, R. Jacob, J. Tibanywana, and G. Makuka. 2018. *In Search of Urban Recreational Ecosystem Services in Dar es Salaam, Tanzania*. Environment for Development Discussion Paper Series EfD DP 18–06. www.rff.org/files/document/file/EfD% 20DP%2018-06.pdf (Accessed 12 September 2018).

Tornaghi, C., and M. Dehaene. 2020. The Prefigurative Power of Urban Political Agroecology: Rethinking the Urbanisms of Agroecological Transitions for Food System Transformation. *Agroecology and Sustainable Food Systems* 44(5): 594–610.

Tripp, A.M. 1997. *Changing the Rules: The politics of Liberalization and the Urban Informal Economy in Tanzania*. Berkeley: University of California Press.

Ventriglia, U. 2002. *Geologia del Territorio del Comune di Roma*. Roma: Amministrazione Provinciale di Roma.

Wang, S., Q. He, H. Ai, Z. Wang, and Q. Zhang. 2013. Pollutant Concentrations and Pollution Loads in Stormwater Runoff from Different Land Uses in Chongqing. *Journal of Environmental Sciences* 25(2): 502–10.

Wegerif, M.C.A., and J.S.C. Wiskerke. 2017. Exploring the Staple Foodscape of Dar es Salaam. *Sustainability* 9: 1081.

Yang, Q., H. Li, and F. Long. 2007. Heavy Metals of Vegetables and Soils of Vegetable Bases in Chongqing, Southwest China. *Environmental Monitoring and Assessment* 130(1): 271–9.

8

CULTIVATING THE CITY FOR ECOSOCIALISM

We have found that urban food-growing plots are highly differentiated in format and purpose, as our site-based studies revealed local and national idiosyncratic features determined largely by historical and contemporary forces. A prominent distinction was the meagre but rising food production in global North cities, while in those in the South output was comparatively high yet stable or in decline. An underlying determinant of this difference is the wide gap in standards of living that exist between the North and the South, and the fact that food is a much larger portion of household budgets in the South—as marketed income as well as cost. A generally unexamined outcome of the world's neighbourhood-sited food growing is its contributions to community organising that produce new social networks serving as frameworks for developing social capital, including learning (e.g. about ecologies) and action (e.g. local leadership). What is missing in this accretion of knowledge and initiative is a strongly progressive political movement. Growing food is not in itself a political act, and social capital can be engaged in neutral or regressive politics. Although just ten in number, our case studies reflected a wide range of biophysical contexts and social and political activity—features that, we concluded, were more significant than food output. Our argument for re-framing urban food growing from agriculture to cultivation is based on these findings. It is the basis for re-directing urban food growing onto a decidedly progressive path—that of ecosocial cultivation.

The urban cultivation sites we examined are distributed throughout the hierarchy of capitalist uneven development, which is a salient feature of today's world economy. We researched sites in five continents, and they bridged the full range of countries, from the richest to the poorest. Using standard (capitalist) national data, GDP per capita (World Bank 2020), the sites are in nations that fall into top, middle, and bottom shares. (The richest, the US, has a GDP per capita 57 times that of the poorest, Tanzania.) In order, the cities located at the top are New York

DOI: 10.4324/9781003131281-8

City and San Francisco, London, and Rome; in the middle, Rosario, Chongqing, Havana, and Potchefstroom; and at the bottom, Tamale and Dar es Salaam. Of course, there are uneven development vectors within as well as between these cities, as capitalism depends upon continual generation of socioeconomic inequality at all levels. The sites represented typical local food-growing schemes that played various social and ecological roles (with varying intensities) in addition to output, including gentrifying neighbourhoods, mobilising communities, developing social capital, and producing sustainably.

Status of urban food growing in the global North and South

Recent revivals in the popularity of urban food production in the most capital-awash cities and more broadly in the global North coincide with shifts towards the rule of financiers, which has decisively remoulded the structure of cities even as some movements (mostly from below, including squatters) have successfully repulsed attempts at transforming their neighbourhoods into capitalists' playgrounds or into (largely white) higher-income households with fungible private property (Busà 2017, 43–5; Sharzer 2012). Since the 1970s, this urban restructuring has been used to widen capital accumulation opportunities through speculation and landlordism, ultimately pricing out poorer inhabitants (a process also known as gentrification). The renewed investment in urban land and buildings has at times been called, euphemistically, revitalisation or urban renewal campaigns. In the prosperous global North cities, where four of our sites were located, there was an internal scale and functional difference based on population density and location.

In cities in the global South, but also in legacy colonial systems in parts of the global North, international investment flows roughly coincide with mass displacement, often by violent means, and countered by intensification in the mobilisation of marginalised communities for the purpose of self-defence and/or struggles for greater autonomy. In other words, more recent mass migrations from the countryside overlap with decolonisation or sovereignty issues. Much of this has been and continues to be closely linked to environmental concerns regarding the protection of forests, agro-diverse smallholdings, water sources, and the like. Outcomes in many global South cities tend to be expansive squatting under parlous conditions in urban peripheries in ever-sharper contrast to fancy, militarily protected urban centre high life. The global South is home to the fastest-growing cities in the world (UN 2019). Of the largest megacities of more than 10 million inhabitants, 75 per cent are in the global South and are expected to grow to 84 per cent by 2030. One of their visible and persistent features is rapid growth of impoverished slums and informal settlements, in which food access and security are habitual problems. At the same time, the expansion of these cities engulfs proximate arable environs.

There are exceptions to such displacement, as always, especially in cities with higher degrees of state involvement in housing and a modicum of welfare provision, as in Chongqing and Havana. Experiences are highly variable, but the overall trend in urban food production is one of steadily waning importance relative to household self-provisioning, as more people are forced into wage dependence. This renders many city dwellers vulnerable to food price oscillations, in part related to the increasing influence of financial speculation on commodified food and closely linked commodities like fossil fuels. Nevertheless, urban cultivation in the global South remains a safety valve for many households, and it can even lead to additional household earnings. However, the incomes are not necessarily redistributed equally among household members, and this may lead to exacerbating economic disparities within households. Again, there are notable exceptions (also depending on how one defines global North and South), such as in Greece (Haniotou and Dalipi 2018) and Russia (Boukharaeva and Marloie 2015; Lemarchand 2018), where gardening has become (in Greek cities) or has been (in Russian suburbs) a means to compensate for health-undermining deprivations imposed by state institutions trying to satiate insatiable capitalists.

In the global North, the trajectory of political economy has been largely one of reinvestment into areas devastated by previous wilful negligence by local governments and through state economic restructuring. Unionised industrial workers were replaced or converted into lower-paid, precarious service and contract workers. Surveillance and militarised policing were intensified in impoverished or marginalised neighbourhoods and is being challenged demonstrably by such movements as BLM. The process has roughly coincided with the spread of environmentalist and localist movements striving for, among many other changes, an ecologically sustainable and more socially just food system (see also Albo 2007; Born and Purcell 2006). As pointed out in Chapter 5, cities where land use is subjected to the highest rent-seeking pressures better reflect the underlying objectives and actual potentials of urban food production.

New York City is one such place, and its WSCG illustrates this convergence among different capitalist and popular demands and their consequences in that it became part of and contributor to a wider gentrification process. In a context of cash-starved municipalities (partly due to deindustrialisation, public asset privatisation, and extensive pauperisation) and a capitalist class flush with cash, gardeners' free labour can be very useful in raising neighbourhood land profitability profile by converting derelict lots into thriving green spaces, concomitant with dislodging the renting poor and welcoming better-off owners. The process, in settler colonial regimes like the US, is deeply racialised and gendered, so that most of the new owners tend to be white, and much of unpaid food-producing work tends to be done by women. The other side of this historical tendency is the rise in urban cultivation linked to increasing levels of penury in neighbourhoods with people of colour or ethnically mixed working-class majorities, as in parts of Detroit, Rome, and Toronto. In those places, urban cultivation has contributed to more food justice, among other positive developments.

Here, the challenges related to intensifying social inequalities have been linked to wider biophysical (ecological and environmental) dynamics. Cities are not just places made by people and with specific social and built environment characteristics.

They are also biophysical systems with many processes that are dialectically related (i.e. they mutually transform each other), within which are also multifarious kinds of social relations. Just as cities are structurally tied to distant places, rural and urban, so do urban ecosystems presume inescapable linkages with other ecosystems and ultimately with a larger planetary reality that features climate change. An unplanted wooded patch, a family of rats in their partly human-facilitated habitats, our intestinal microflora, among much else that is non-human biota, negate any easy separation between us, other organisms, and our shared physical settings. This is not to pretend any equivalent causality or influence of non-human and human processes and relations in creating and transforming urban environments. Causal primacy and determinative influence may rest unequivocally in how societies are structured and organised while differences among urban ecosystems are only partly derived or decided societally.

The interplay between the biophysical and social becomes especially obvious when food is grown in urban areas. Regional climate variability, pre-existing landforms, co-evolving relations between different species, among other factors, all affect, alongside social struggles and technical interventions, specifically where cultivation is viable—for example, in determining what crops can be produced in what quantities. Even in the most obvious cases of human impacts, like soil pollution and the urban heat island effect, there are always fingerprints of physical processes and non-human organisms. The heat island effect and toxin-laden dust may be locally lessened in a regional climate characterised by a predominance of lower temperatures and strong winds, or by canopies of pre-existing forests or by fresh tree species establishing themselves because of favourable habitats created by human impacts. Microbial species and intrinsic soil properties may stifle the mobility of pollutants into vegetables as a result of building debris. These scenarios not only show why the ecosocial dynamics of food growing are not necessarily predictable (and not just in cities). This is gleaned at all scales of analysis, within and beyond cities, where global climate change affects prospects for urban food production differentially. The threats include recurring severe droughts and floods, among others. These grand contours may be remote and unhelpful when attempting to produce food in a city, but they point to the importance of finding out about and facing both social and biophysical challenges that are specific to a place, while also linked to and influenced by what happens beyond that place.

In part, this is why urban cultivation does not necessarily bring about net ecological benefits. It is positively impacting for some organisms more than others, while the greater biodiversity achieved is not necessarily inclusive of native species (which may lose out). Carbon sequestration, often cited as a major benefit, hinges on food production practices. The benefit is not a given, especially compared to other kinds of green spaces with large patches of grass or, even more effectively, trees (depending on the kinds of trees planted). In like manner, local production that obviates fossil-fuelled transport is not an ecological sustainability given; this requires comparative technical assessments of local and distantly sourced crops. The least controversial biophysical effects of urban food production would be in improving soil quality (or in making new soil) and reducing soil erosion. However, these outcomes depend on where the gardens are, what people do in those spaces (if, for part of the year, the soil is devoid of vegetation, soil erosion may result), and how extensive gardens are relative to paved area. The attenuation of urban heat island effects, while possible by growing food, also depends on how large an area is cultivated and where. It might make little to no difference overall for a city if the cultivated land is very small compared to total heat dissipation and built-up surface. In sum, urban food production may not yield any more ecological improvements than green spaces in general, and perhaps cultivating urban land can be less effective towards attaining ecological sustainability, which is arguably impossible without simultaneously undoing and overcoming the town–country separation.

All these considerations, discussed in the previous chapters, indicate that conventional claims about urban agriculture grossly exaggerate potentials for food productivity and ecological sustainability. In the celebratory and Panglossian view of foodist and localist worlds, social injustices and environmental challenges are downplayed, if not swept under a rug of individualistic do-goodism or can-do technocratic ingenuity. To us and many others who have been critically re-evaluating urban agriculture, its social effects (including on human health) and political consequences (including on inequality) have the greater upside potential. We therefore re-frame the production of food in cities from agriculture to cultivation. This is to underline its social processes and contributions and to distance it from the unreasonable expectations (often market-reductionist) implied by the term agriculture.

We find especially perplexing claims that producing food in cities will help reduce hunger. This flies in the face of manifest and consistent failures in securing food access to all irrespective of technical achievements that have led to unprecedented rates and amounts of food output (and waste). The precarious situation of many urban food gardens and, especially in highly capitalised cities, the recurring use of plots for gentrification projects, should make it obvious that technical solutions are imbricated in political objectives largely inimical to the well-being of most people. At the end of the day, there are no politically neutral urban designs or plans or other kinds of technical and administrative interventions. This is for the basic reason that designating land for particular uses is already a process of deciding who is to live and do what, where, and how; that is, it is already political by definition. Institutionalised expertise is insufficient and even dangerous when political implications and wider social consequences remain unaddressed. The various neo-localist and foodist fashions of the day, especially among the privileged chattering strata, reinforce a capitalist logic of urban agriculture.

Building on existing accomplishments and approaches

What urban cultivation can achieve instead of agriculture, as many urban cultivators already understand, is greater social sustainability in cities and (at best) a stop-gap measure for communities suffering from the imposition of food shortages or nutritional inadequacies. The social sustainability potential is especially useful towards community development, including environmental education and public health promotion, as explained in Chapter 5. The social benefits can be reached provided that biophysical processes are addressed to the same degree of political conscientiousness. Attending to only one or another aspect is insufficient in addressing the twin crises of climate change and increasing inequality. In Chapter 7, we illustrated with case studies how ecological consequences of urban food growing can be problematic. What is also worrisome is the exposure to contaminants that urban cultivators may be facing as a result of their marginal position, and lack of support to assess levels of risk and prevent exposure. The result may be a hidden form of environmental injustice yet to be appreciated and studied. Only in the case of Havana, among the localities examined here, is there a concerted effort, involving institutions and grassroots initiatives, to address social as well as ecological challenges simultaneously by prioritising both food access and agroecological techniques. Physical environmental problems, such as soil and water contamination, have also received attention, but without yet being met with commensurate measures.

There is much that can be drawn from the Cuban experiences, but also from widespread gardeners' collective efforts demonstrated in the many studies so far carried out in other countries—in both the global North and South. Plenty of remarkable work by many urban cultivators and allied activists (including some scientists) has been done to identify socially promising practices and situations, to improve food production awareness and capacity as well as food-growing strategies, and to develop and promote safeguards against health risks. The development and diffusion of cultivation skills in cities is already a major contribution that helps overcome various forms of capitalist alienation processes, including the urban–rural and nature–society separations. Having a personal connection to the full lifespan of just some of one's food consumption can be an enlightening behaviour-challenging process for urbanites in today's world, creating the possibility of mental and physical health-promoting rewards, as well as social ones.

Above all, urban cultivation, if carried out as a communal or collectivist project, enables people to get to know each other more substantively and, thereby, to establish reciprocal trust and develop new or reinforce existing practices of mutual aid (a sort of resource redistribution external and alternative to formal institutions and capitalist markets). The emphasis here is on communalism, based on collective undertakings. Gauging from the numerous case studies available, the vast majority of urban cultivation projects do not involve turning urban land and instruments of production into commons, that is to say, subsistence-oriented sharing and use of resources according to community-based rules established and modified by consensual means (Clark 2016). These are the kinds of projects that transcend the political limits of individualistic or household-delimited cultivation. Urban community gardens that amount to internal allotments of plots to individuals or individual households are, from this perspective, rather limited in political scope (Tornaghi 2017). However, the sharing and collective management of the means of cultivation and uncultivated spaces do contribute to building relations based on mutual aid that results in self-managed (some say autonomous) urban space outside capitalist and statist logic—at times with at the expressed aims of "retaking the city" (Mudu and Marini 2018, 3–4).

Such patterns of mutual aid are visible in Rome's squats and some urban community gardens in London and New York City, for example. In South Africa, the development of urban food growing is directly related to resistance to ongoing neoliberal exclusion rooted in the apartheid period (Siebert 2020). Some of the more collectivist cases of urban food production, as in Chongqing and Havana, demonstrate the possibilities and limits of taking over urban spaces for cultivation. In Chongqing, cultivated spaces are established and run collectively by squatters but production occurs in household or individually managed plots, in ways resembling some forms of community and allotment gardens in global North cities. In Havana, institutional mandates and publicly furnished means of cultivation favour a mixture of individual and household initiatives as well as wider, largely grassroots-based, community-controlled cultivation spaces (*organopónicos*), but they are contingent to a major extent on national state interests and the wider geopolitical conjuncture. All these variegated experiences contain a germ of urban communalism and the potential for practical alternatives to capitalist ways of running a city.

These aims can also be achieved in other ways, which ideally should be linked up to develop shared objectives for wider application. Think of socially constructive interactions and community-building processes in public playgrounds and other green spaces, for example. What is specifically unique to communal forms of urban cultivation is that they engender and nurture the kinds of collective self-management skills that are useful to producing spaces in directly participatory ways and for taking greater control of one's lived urban space. Community gardens not only strengthen people's social capital, they can become a fulcrum for prefigurative explorations of planning from below, demonstrating that people can and do build and run spaces and can do so on a communal basis.

Many squats worldwide (and, in the Americas, historical Maroon communities in the countryside) have shown the viability of collective self-management (Martínez López 2018; Roberts 2015; Shoatz 2013). Urban community gardens can add to such long-existing strategies for creating spaces in ways that each contributes according to ability and ensures everyone's needs are more or less met. However, without wide and sustained external social support and coordination of joint actions for political change (from neighbourhood to world scale), such everyday productions of lived space tend to fade, or be co-opted institutionally, or be violently removed. For progress to develop, we need an overarching political project that appeals to and engages peoples' minds and labours. This is not at all to suggest that such attempts have not existed or that everyone must make their projects conform to some singular final objective and set of political strategies. For the case of urban cultivation, however, it seems no such discussion on coordination has occurred at all and is long overdue. So, we recommend that the disparate ideals represented by urban cultivation projects should be harmonised by seeking and stressing overlaps, common understandings and objectives, and by developing strategies that are not mutually incompatible and are sensitive to situations (unity in difference).

To do so, aside from first securing urban space access and communal control for the long term, requires raising awareness of such a need, setting up the means for continuous dialogue among different urban cultivation communities (and beyond), establishing mutual trust and mutually agreed procedures, doing outreach to gain the widest social support (beyond single cities), and coordinating plans of action across contexts. It is beyond this work to spell out a political programme or platform, and we make no claim to providing any ready-made practicable solution or to any novelty. There are, in any case, many who are more competent than us, including those who have been formulating innovative programmes and platforms and even putting them into practice. We currently find especially compelling examples in Cooperation Jackson (Akuno, Nangwaya, and Jackson 2017), Zapatismo (Subcomandante Marcos and Le Bot 1997; Baronnet, Mora Bayo, and Stahler-Sholk 2011), and the Rojava insurgency (even if now highly compromised by unsavoury alliances precipitated by the need of sheer survival), as communities with sufficiently similar ultimate objectives while employing strategies that are very different because they conform to specific local circumstances (see also Adamovsky et al. 2011). We think that such examples furnish a general framework for placing urban cultivation within a socially much larger political project, aiming to instate ecologically sensible and egalitarian relations. For us, this much larger project is ecosocialism (discussed later; see also Chapter 1). It has been recently embraced by Cooperation Jackson in Mississippi, and aspects of it have been adopted officially or otherwise by various political parties, e.g. the US Green Party, the European Union's Nordic Green Left Alliance, and the Socialism and Liberty Party in Brazil (for an overview, see Engel-Di Mauro forthcoming).

While all such direct and indirect contributions and struggles are of decisive importance, the challenges of specifically urban biophysical conditions, especially pollution and climate change, remain salient and are not resolvable by focusing only on changing social relations. Such challenges bring into relief the need for methodical approaches that can help find alternative practices to enable cultivation in ways that are public health-promoting and ecologically sustainable. This cannot be expected from urban cultivators alone, who may or may not have the background to evaluate biophysical processes and their health hazards. Moreover, the field and laboratory instrumentation and infrastructure required surpasses the wherewithal of most community cultivators, especially those who stand against or are independent of formal institutions. This means reliance on scientists with appropriate expertise and with access to the necessary resources to carry out evaluations. In itself, this need not be politically compromising or undermining. Much ammunition for anti-capitalist environmentalism emerges out of findings based on conventional scientific approaches and through mostly state institutions. The movements for climate justice are one salient modern example, as, irrespective of their politics, climatologists and like scientists have been largely those identifying the problem and its causal mechanisms. They has been a principal knowledge resource in the case of urban cultivation, as pointed out in the preceding chapter, with potential to promote templates for an ecosocialist practice in cities, depending on how that knowledge is used.

At the same time, as many Marxist, socialist feminist, and other critical scholars of science have pointed out over decades, mainstream science cannot be taken at face value. We should add that current popular denials of science, especially as it relates to public health issues (e.g. long-standing anti-fluoridation and anti-vaccination sentiments in the US) as well as opposition to social practices (such as wearing masks) to contain the spread of COVID-19 in 2020, are not supported by our critique of science, which centres on its reform, not its rejection. Our critique supports a "science for the people," not one for capitalist profit (see www.scienceforthepeople.com).

There have been and there are highly problematic assumptions and practices associated with institutional science, related especially to its financial support (and guidance) by corporate capitalist enterprises. One need only recall eugenics as support of racist notions by means of scientific argument (see DenHoed 2016). In the case of urban cultivation, the matter may be more subtle than that, and we have pointed out by means of examples (Chapter 6) the sort of political biases hidden behind technical discussions. For urban cultivation to steer society away from capitalism and towards Ecosocialism (or something like it), there are not only many combined practical social and biophysical challenges to face, but also ideological ones related to the false, capitalism-promoting, or capitalism-friendly solutions suffusing much of urban food agriculture advocacy.

The failures of reactionary populist Brexit and Trumpist regimes to deal with the COVID-19 pandemic illustrate the extreme negligence of two capitalist kingpin nations to support public health science and practice. Thus, by 10 August 2020, the UK and the US had together suffered 29 per cent of COVID-19 deaths while having only 5 per cent of world population (Elflein 2020). The COVID-19 pandemic (CDC 2020; WHO 2020) and the global ballooning of the BLM movement (Buchanan, Bui, and Patel 2020; Chotiner 2020), as we noted in Chapters 3 and 5, beckon some special regard, as they constitute a major conjuncture within which this work is situated, even if we started writing it prior. Our broad ecosocialist perspective on these events includes the following points:

(1) Food production and consumption today are increasingly dependent on global supply chains that are corporate based. Adequate public health monitoring and financing are not part of such capitalist systems as their costs subtract from, rather than advance, enterprise profitability. Consequently, public health programmes that are bedrocks of social sustainability have not been in the least prepared to identify and contain COVID-19-scale threats.
(2) After COVID-19's emergence and spread, advanced capitalist economies like the UK and US reacted by a singular focus on restoring their functioning via a rapid V-shaped market recovery. This meant interruptions to necessary virus

detection and prevention practices. The resultant duel between saving capitalism (the "economy") and promoting public health prolonged the pandemic, costing many more human lives. In less than 1 year, the US suffered (through 20 August) 63 per cent more deaths than it did from its four interventionist wars in Korea, Vietnam, Iraq, and Afghanistan combined.

(3) COVID-19 raised widespread public concern about the fragility of food provision: "There's nothing like the sight of stripped grocery store shelves to focus people's attention on where their food comes from" (Mark 2020). One result has been a renewed effort to grow food in domestic and community gardens (Wharton 2020; Scott 2020); these have been tagged as COVID-19 victory gardens, recalling WW II (Weinberger 2020). As we have pointed out, any attempt at urban self-sufficiency in food is bound to fail in most cases, so food provisioning must involve guaranteed food access to all and development of mutual support networks transcending the city–country divide, building on linkages between cities and peri-urban areas. In some respects, mutual aid redistribution networks established largely by grassroots initiatives (with important anarchist involvement) prefigure such an ultimate objective, even if they are not currently leading to any appropriation of the means of food production or an integration with wider food production systems. Some examples of these initiatives can be found in the Boston area, US (Massachusetts Jobs with Justice 2020), and Brighton and Hove, UK (https://brighton-mutualaid.co.uk/).

(4) COVID-19 furthers the racial discrimination that is an institutional legacy in the US (and other nations as well). People of colour are disproportionately infected as well as killed by the virus, a reflection in part of government negligence in public health efforts (Selden and Berdahl 2020). This parallels the racialised classism pervading current urban food production. However, without full, universal healthcare coverage that also focuses on undoing historical discrimination practices, racially minoritised communities will be decimated by illness, undercutting their food procurement and production capacities and reinforcing their dependence on capitalist institutions.

The connections between cities, food, and viral pandemics are certainly complex, but capitalist economies have a record of miserable failures in times of public health crises. The struggles for public health and against its racial inequalities would command leading roles in establishing ecosocial programmes for urban food cultivation.

Some technical experts take for granted prevailing social relations and ideologies and thereby side, by default, with the position of those with greater political power. This means that one must not only be aware of biophysical challenges and repercussions to cultivating in cities, but also become discerning of its internalised political biases or tacit allegiances (for examples in the soil sciences, see Engel-Di Mauro 2014). Expert recommendations, in other words, can be highly problematic when it comes to assumptions about what goes on in society or how society ought to be organised. Many technical solutions, paraded as politically neutral (and without biophysical assessments), are, on offer, aiming to curtail output disadvantages of urban agriculture through capitalisation (e.g. to building high-rise farms that market food). Such business solutions must be challenged if urban cultivation is to serve socially and ecologically constructive ends.

Adding to the reactionary ideology hidden in business approaches, there are also technical flaws reflected in the underestimation, inconsistency, and/or confusion about contamination processes and remedies. There are numerous ways urban cultivators and the consumers of their produce can protect themselves from contaminants, but the prevailing technical frameworks prove to be incoherent or muddled, and even more so among encomiastic promoters of urban agriculture. Two major flaws in the current discourse illustrate this. One is the widespread misunderstanding of contamination processes, lacking appreciation or even awareness of the major challenges they present. Another emerges from a largely disjointed discourse that in part follows from an inadequate or compartmentalised grasp of contamination issues. This leads to contradictory recommendations about contaminant containment and its role in urban food production. Both flaws are fundamentally political in character. Yet, engaging scientists who are knowledgeable about and open to progressive alternatives is not easy. In conventional liberal thought and especially in the biophysical sciences, scientists may be regarded as biased if they question or expose the biases of capitalist ideology in their work. Rachel Carson, a marine biologist (see Hecht 2012), and Stephen Jay Gould, a palaeontologist, for instance, were stand-out scientists whose reputations were attacked by peers on such grounds.

Apart from such political considerations, there is another hurdle to overcome in developing an ecosocially mobilised version of urban cultivation. In addition to (typically dissembled) political divides in the sciences, there are extensive gaps between people with practical applied expertise and those with scientific technical knowledge. With regard to urban cultivation, most cultivators do not have the sort of background that enables them to understand the terms and processes used in training biophysical researchers. Conversely, most scientists do not have cultivators' expertise regarding local relations of power, how to produce and process food, and how to manage a community garden.

This divide creates another obstacle to developing and implementing ecologically sustainable cultivation practices and health safety measures. Scientists, at minimum, must recognise their limits and find ways to understand the specific social situations of urban food growing and to be open to suggestions for, and critiques of, their work. When such socially responsive scientists are too few or lack institutional support, as is currently the case, it becomes difficult to attend to research translation and explanation for wider diffusion, much less to provide accessible training to cultivators or the public at large. This is another reason that the Cuban example is of importance in promoting an ecosocialist urban cultivation, irrespective of whether one agrees or not with state-centred forms of socialism. Having unique and strong institutional support is beneficial not only towards developing and spreading ecologically sustainable urban food-growing methods, but also in serving as an example to the rest of the world with respect to what is technically and practically feasible and positively transformative (see also Hearn 2010, 179–80). A unity in diversity when it comes to political strategy entails, in the current context, reconciling diverse forms of engagement, ranging from conventional political party formations and state structures to initiatives for (already established) communalistic autonomy and confederalism from below. This, we recognise, is no easy task, and there are many contradictions to surmount that can explode as open and deadly hostilities. The risks are great, but we argue that they need to be weighed against capitalism's record of planetary-wide ecological and social catastrophes.

Assembling ecosocialism

The manifold practicable and practised egalitarian alternatives to capitalist relations described earlier overlap with and are necessary towards building ecosocialist communities that would make of urban cultivation a set of ecologically sustainable practices integrated with other facets of life, within and beyond cities, and devoted to the purpose of feeding people. Before discussing the contours of its food production in cities, a fuller, though abbreviated, explanation of ecosocialism is in order (see also Chapter 1). Generally, an ecosocialist position resembles much of the current anti-authoritarian left politics in that capitalism is understood as intrinsically destructive of society and environment. What sets us apart is the view that an ecocentric remaking of society is as important as establishing a world based on equality (not to be confused with sameness, as reactionaries are fond to do), as outlined by several ecofeminists (e.g. Mies and Bennholdt-Thomsen 1999). Caring, sharing, and other mutually supportive action (i.e. mutual aid) characterise ecosocialist practise, but competition, selfishness, and related individualistic or potentially harmful tendencies are not suppressed. Rather, they are rechannelled to more constructive ends. An example is the establishment of rewards for those competing for most gift-giving without provoking self-harm or feelings of indebtedness in others, as customary in societies like the "potlatch" ("to give", in Chinook) among the Kwakwa̱ka̱'wakw of Northwest North America.

Ecosocialism thus precludes any space for profit-oriented activity, since endless capital accumulation is a main feature of capitalist relations. Profit is here understood as appropriating others' labour for the sake of wealth accumulation by the owners of land and capital. Leftists concerned with farming and food systems would do well to return to Marx's (1992, 694–5) analysis of the origins of the capitalist farmer, where it is made clear the issue is not the form but the mode of production that regurgitates the transformation of a few smallholders into ever-larger capitalist landowners or, in our current iteration of capitalism, into the next global agribusiness. Cargill, currently the world's largest, is responsible for about one-fourth of US grain and meat exports (Sekulich 2019). Urban cultivation is hardly immune to the reach of such a general process.

Ecosocialism is ecocentric, which refers to structuring our ways of life such that ecosystems are not endangered by our striving to ensure that every human being can achieve full self-development—without hampering the self-development of others. To some, this may seem at most as a weak ecocentrism or more like a kind of anthropocentrism (see Pepper 2003, 33). We understand that preserving ecosystems as they are is not the aim. In fact, trying to do so can court calamity, like suppressing regular low-intensity fires that help maintain certain ecosystems and enable them to evolve (i.e. for many of the species in such ecosystems to co-evolve). Rather, as many ecologists have realised and as many Indigenous peoples have long understood, the issue is one of preparing for a range of possible changes known from long-term observation or study. The short-sightedness and predictability-obsession intrinsic to capitalism-infected science is a major impediment to overcome.

Within this overarching aim, therefore, most other organisms have to be included in a mutually beneficial co-evolution, since we make ecosystems with them. To be more precise, we are thinking of those beings that are not lethal to us, provided that those harmful to us are not also keystone species whose disappearance would provoke a cascading effect leading to other species disappearing and the sort of ecosystem instability that undermines our lives as well. In such cases, containment for self-defence would be appropriate. What is more, ecosocialists also often stress that the struggle to forge ecocentric societies is, at the same time, a way of avoiding the increasingly likely prospect of a threat to humanity (possibly via global climate change) as well as a way of stopping the continuing human-induced extinctions of other species (Kovel 2014; Schwartzman 2009). Ecosocialism is a way of combining thought and action on the basis of ecological principles (broadly understood, inclusive of knowledge systems where nature and culture are not split) as well as Marxist insights on capitalism, shed of their commodity productivism (Dickens 1992; Kovel and Löwy 2001; Löwy 2011, 31–2; O'Connor 1988).

Ecosocialism fuses, or is striving to merge, diverse yet potentially convergent perspectives that have regrettably never been joined or have come about through historical schisms. These may be variously called anarchism, autonomism, communalism, communism, collectivism, democratic confederalism, etc., according to the many currents that have close affinity or overlap with Ecosocialism (see also Pepper 2003, particularly Chapter 5). A major current is also indigenist, which is an abbreviation alluding to the thousands of state-free peoples who already live communalistically and in ecocentric ways. For such societies, the matter is not achieving an end, but resisting impositions of change and retaining traditions as well as struggling to revive them. However, not all traditions may be compatible with ecosocialist principles (e.g. their patriarchal norms, gender binaries). Whether or not all such currents and Indigenous lifeways can be finally reconciled is a work in progress and a political priority for ecosocialists. Ecosocialism may be internally diverse, but under the broader umbrella of socialism there are certain unifying principles that demarcate it from other currents that have adopted environmental concerns.

First and foremost is the desire for an ecologically sustainable *and* egalitarian society organised as freely associating people sharing the means of production, with contributions offered according to relative ability and allocations, accepted in accordance with changing needs (Kovel 2007, 243; Löwy 2005; Mies and Bennholdt-Thomsen 1999; Pepper 2003; Salleh 2009).

Secondly, ecosocialism is firmly anchored in perspectives originating from intellectual elaborations derived and experiences learned from oppressed peoples, coupled with Marxist and allied materialist-oriented theory and methodology. This is why ecosocialism can cohere with multiple forms of socialist approaches, including (eco)feminist, decolonisationist, anti-racist, anarchist or left (libertarian) communist, and left (red–green) environmentalist frameworks (in this, we disagree with the more pessimistic view in Pepper 2003). These all help identify processes intrinsic to a capitalist mode of production and demonstrate how capitalist relations have global-scale ecologically degrading repercussions. There is no ecosocialism without a focus on and struggle against all forms of social relations of domination and their linkages to the treatment of the rest of nature (Mies and Shiva 1993; Salleh 1997; Turner and Brownhill 2004). This is why materialist ecofeminism can be thought of as a principal pillar (Kovel 2005) as well as many Indigenous peoples' worldviews and practices. The latter is particularly important towards purging the ethnocentric (including Eurocentric) prejudices and teleological tendencies characterising much of historical Marxism and socialism more broadly, which is intimately related to commodity productivism (Bedford and Irving-Stephens 2000; Forsyth 1992; Kovel 1991).

Third, given its Marxist underpinning, an ecosocialist approach includes historical and dialectical materialism, which is, by definition, a non-deterministic and open-ended method (Clark 2014; Engel-Di Mauro 2017; Kovel 1995, 38–41; Levins and Lewontin 1985; Löwy 2011; Paolucci 2007; Wan 2012). This implies that there is no final endpoint to history, social and ecological. Crucially, historical and dialectical materialism has a basis in praxis, which is the mutually transforming relationship between thought and action, theory and practice, means and ends (Kovel 1998).

Fourth, ecosocialism will flounder without a scientific basis. By this we mean science that includes urban cultivators' and Indigenous peoples' modes of knowledge systematisation. It is also a science that is self-consciously critical of, if not entirely free from, capitalist influence (Engel-Di Mauro 2014; Levins and Lewontin 2007; Schwartzman 1996)

The main strands of socialism (in the widest sense) that have contributed to the building of ecosocialist perspectives are many, but to some extent they all share much with Marxist understandings and method. Drawing from many socialist and oppressed peoples' perspectives and environmentalist thoughts, one can trace the relationship between capitalist relations and the current global ecological crisis, and on such basis understand what steps must be taken.

We as ecosocialists, in common with many socialists, understand capitalism as a class-differentiated mode of production underpinned by multiple forms of oppression, in which the fruits of people's labour are appropriated by those controlling the means of production. How this occurs, develops, and is maintained depends on historical context and is place-specific. Crucially, classes are never unitary or stable, being forged through a multitude of social relations of power, and by the splitting and fragmentation of working classes along gender, racialisations, sexuality, age, presumed abilities, and so on (see also Marx 1959, 1992). There are some general ways whereby capitalist relations become preponderant: by violent and even genocidal means justified through destructive cultural constructs (sexist, racist, homophobic); class stratification through acts of enclosure (e.g. wresting the means to life away from communities), privatisation (i.e. depriving the majority of society to benefit a handful), and ideological inculcation to normalise culturally these processes.

In capitalism, people's abilities, entire peoples, non-human organisms, ecosystems, and physical environments (now even outer space) are treated as things whose worth is ultimately decided by means of market exchange, rather than on the basis of usefulness or social needs—much less any intrinsic value, that is, value not directly linked to society, like dead fungal hyphae in soils that enhance soil biodiversity and overall soil fertility. Life's realities become reducible to exchange value, and one's labour can become worthless if not performed for market exchange (e.g. for wages). A few rack up modest to enormous fortunes while most struggle to see some improvements in living standards, or barely survive, or perish altogether. Those most successful in this socially and ecologically selective system of depredation (i.e. in sum, globally devastating) manage to take part in the appropriation or extraction of surplus (e.g. profit) from others and to reinvest part of it in activities generating more surplus (this has been done by a great variety of techniques, from slavery to financial derivatives). In this manner, the wealth produced by society as a whole is pocketed or otherwise controlled by the capitalist few, while the destructiveness of constant surplus production is distributed unevenly across society (otherwise put, profits are privatised and costs are socialised). Part of this wealth is directed towards political influence and control, or plutarchical rule. The coercion of human endeavour into market exchangeability simultaneously produces people's historical and progressive alienation from the products of their labour, from the act of production, from their species-being as part of nature, and from each other (Gare 2016; Marx 1978).

The overarching impetus for these ecologically and socially degrading and destructive processes lies in the systematic compulsion to accumulate capital indefinitely. "Enough" is an unused word in the vocabulary of capitalist aggrandisement. Failure to maintain profitability leads to the inexorable extinction of a business and its wage work, so everyone must constantly keep up profit rates (i.e. extract surplus) and seek ever-higher wages. The penalty is business losses or closure for capitalists and poverty for many. Some find alternative means of well-being or survival (e.g. farming on squatted land, violent re-appropriation, like brigandage, etc.) or join (or join with others to form) communities that are not linked to capitalist economies. This latter alternative is part of what ecosocialism is about.

Endless accumulations of capital and processes of alienation bring forth and nourish commodity productivism and its corollary, consumerism, leading to unprecedented social and ecological devastation at the planetary scale. Plastic pollution is becoming its contemporary global icon. Interventions (usually by the state or social movements) that raise the costs of profits and destruction, such as effectively enforced regulation and successful legal action, help alleviate the problems, but only temporarily (O'Connor 1988; Wallerstein 1999). They cannot reinstate who and what have been destroyed; genocides and soil removals are irreversible.

For an ecosocialist form of food production in the city

Our goal in this volume has been to broaden the conceptual frame for city food growing by including assessments of its productivity and its social and ecological sustainability and then developing an alternative analytical framework. Our ecosocial approach is grounded in historical and dialectical materialism that understands the social and the ecological (as Marx and Engels understood long ago; see Dickens 1992). We argue that urban cultivation (rather than agriculture) can play a role in the development of sustainable cities, but that role needs to be empirically specified. In addition to the critique that urban food growing has been used as a basis for the gentrification of city cores after deindustrialisation, and that it is marked by internal contradictions, urban cultivation presents other problems that remain inadequately addressed: it has no realistic potential for producing more than meagre food stocks for urbanites, and its relationship to ecological sustainability has delimiting liabilities. However, beyond food output and ecological sustainability, urban food growing demonstrates ample potential for significant contributions to social sustainability, especially in supporting environmental educations and public health schemes.

Ultimately, for urban food growing to move beyond its oversold vision of producing large quantities, it will need a twin focus on society and biophysical conditions, as well as on political mobilisations that challenge bourgeois food-ism and the neoliberal shoring up of the increasing economic inequalities it is supporting (Sharzer 2012). Research indicates to us that to this point urban food growing cannot reach the level of agriculture, which is generally recognised as based in mono field crops (grains) and husbandry (non-human animals). Urban food growing has no potential for producing field crops, *some* potential for producing fruits and vegetables, and a very low potential for meat production. To assess a realistic role for urban cultivation, we should be looking at food provision differently. If the food system is recognised as involving produce supplied by oligopolistic intermediaries (retailers) from ever-more consolidated primary producers (industrial-scale farmers), many parallels with energy generation and distribution become apparent, suggesting a need for reforms to promote sustainability and, in particular, resilience to climate change impacts (Smith et al. 2016). Urban cultivation is a local phenomenon, the base level in a food system. Its unique grassroots activities contain and incubate adaptive, flexible possibilities for environmental justice and public health whose effects can extend throughout food systems. A progressive politicisation of food growing will need coordinated and joint efforts from the local through the global levels.

The hope that urban food production might produce enough food to support the population within its borders is unrealistic if not delusional. The greatest opportunity for urban areas to reach a higher level of food security lies in the next tier of available land that is beyond the urban periphery: the broader region that is still largely rural. For example, in 2009–2010, 57 per cent of London's consumption of fruit and vegetables was grown in the rural hinterland beyond its urban and peri-urban zones (Growing Communities 2012). However, even assuming a full development of the broader foodshed region, it will still be necessary to bring in food, including cereal grains and exotic foods, most of which are internationally traded. For example, many countries of the global North have long lists of imported exotics, headed by bananas, citruses, coffee beans, and tea leaves.

While groups of our ancient ancestors ate only what they could find by walking within a gatherer-hunter food system and without major intragroup differences in food access, we, in the present, live in a predominantly capitalist system where global-range appetites are met by huge mining and other extractive operations, industrial-scale production, long-distance transport of agricultural products, and distribution that features massive and widening inequality. Because of structural factors, this general picture will not be changed by promoting local food production, but this does not mean it should be abandoned. We think it can still go a long way in promoting the cultivation of environmental and social justice, basic building blocks for ecosocialist futures.

Urban cultivation may be compatible with many kinds of society, including capitalist ones, but we see much ecosocialist potential as well from producing food in all cities. As instances of community-based planning, urban food growing projects disrupt the typical authoritarian character of organising and structuring cities and the tendencies to chase non-human beings away or to keep them hidden from view, often even violently by means of life-destroying chemicals. There are two simultaneous aspects to this. One is the striving to confine and control life (both human and non-human) that impedes capital accumulation—persistent in cities in spite of countervailing ideas and practices. Part of this control is expressed in the common refusal to understand cities as part of nature, as ecosystems in their own right. Another is by splitting cities from their mainstay food sources, which reside elsewhere, often maladroitly referred to as the countryside. This tendency to deny the obvious and to impose a brutal capitalist "development" through, for example, eviction of people from their homes to enable deforestation and resource extraction needed to build city infrastructures. These have been historical products of capitalist relations. A result is that we are alienated in multiple ways from nature (including from ourselves) and hence the very sources of our being and our sustenance (Marx 1978, 1992, 173).

Urban cultivation might become a way to counter alienation as part of wider efforts at overcoming capitalism, since it means knowing and acknowledging our natural roots and the nature–society split characteristic of capitalism. When urban food growing gets in the way of profit, its existence is called into question, and it is often replaced with more compliant projects. This indicates their potential to clash directly with capitalist-favoured re-orderings of urban spaces. In this sense, even as community gardening has been rightfully critiqued by some for underwriting neoliberal capitalism, community gardening remains conducive to resisting this latest form of capitalism, such as by promoting the use value and de-commodification of food (Barron 2016).

However, it must be understood that socially constructive contributions through community gardening may be temporary, resulting from the prevailing balance of forces. Without securing communal control over the means of food production (including distribution and much else) and coordination with communities elsewhere, community gardening will not overcome capitalist relations. Typical even of many leftist promoters of urban food production is the inability to grasp the multiple forms of alienation intrinsic to capitalist cities and capitalist relations generally. It is for this reason that some avowedly leftist scholars of various theoretical stripes, such as urban political ecology, can expect the structural entrenchment of alienation from one's food supply to be overcome simply by commercialising city-grown food (e.g. Bellwood-Howard and Bogweh Nchanji 2017). Nevertheless, it would be counterproductive for us to equate struggles for social equality with the stimulation of a sense of wonderment or re-enchantment with "nature" in the city. It cannot be sufficiently stressed that positive (progressive) environmental or ecological outcomes are not necessarily tied to particular political objectives. They can be achieved by autocratic authoritarian as well as by democratic egalitarian means, through mass evictions and individual private ownership as well as by inclusive mutual aid processes and communal property. There is no necessary direct correspondence between what happens in ecosystems and what happens within just one of their components (Haila and Levins 1992).

The potentials of urban food growing therefore lie primarily in its social output and in its (at best) inchoate questioning of capitalist ideologies. A transition to ecologically sustainable food access depends on surmounting numerous difficult challenges. One is overcoming prevailing property relations. This includes combating incoherent or plainly incorrect notions or framings about the commons, which some reduce to the mere sharing of urban green spaces (Borch and Kornberger 2015; Dellenbaugh et al. 2015) and some others, incredibly, conflate with municipal property (Colding et al. 2013). Such conceptualisations are politically pernicious in that they re-direct attention away from the vast power inequalities involved in who ultimately calls the shots in a city. In contrast, Cooperation Jackson (Akuno 2018) and several collectives in Rome (Mudu and Marini 2018) employ urban community gardens as pre-emptive strategies to curtail land speculation. This offers an important way of calling capitalist property systems into question and establishing pockets of actual urban commons. However, they can only constitute the first steps towards setting up actual communal property based on social equality. Revivalist community gardens demonstrate a prime example of the complexities of contemporary urban enclosures or commons (Eidelman and Safransky 2020). As ecosocialists, we argue for a re-framing of the property relations that are involved in the local regulation of urban cultivation, often located in what are described as "commons." In fact, the relations are top-down, dominated by private (profiteering) and state (taxing) interests, rather than bottom-up, built through practices that emerge in neighbourhood sites (Ela 2006).

The present capitalist food systems require structural changes in order to promote ecological sustainability (Martin, Clift, and Christie 2016). The matrix of challenges is daunting: (1) reducing waste, which accounts for up to a third of production through present food chains (Kummu et al. 2012); (2) shifting crops away from animal feeds and biofuels to human foods, which can increase global calorie availability by up to 70 per cent (Cassidy et al. 2013); (3) shifting to sustainable and healthy plates on the consumption side (Macdiarmid et al. 2014; Sage 2012; Thompson 2013); (4) adopting food-growing intensification practices in which productivity is raised without increasing environmental impact and without using more land (Garnett and Godfray 2012); and (5) carefully studying each urban situation as sets of biophysical relations to enable the development of mutually beneficial relations among species, including us, to reduce exposure to toxins and other hazards, and to prevent further pollution. Cuba appears to have made some progress in at least the first four aspects of this matrix.

The ecologically sustainability dimension implies that scientific, technical expertise must be undertaken as an inextricable part of social justice struggles. Under capitalist conditions, the wealth of scientific knowledge production is largely inaccessible to most of those who would benefit from it. There are institutional obstacles to generating support for studies that are accessible to society at large. Were this the only problem, one could argue, as many capitalism-friendly reformers do, for improving education (e.g. scientific literacy), access to scientific works, and the like. But this does precious little to make for socially responsible and responsive sciences, as pointed out earlier. What is needed is a much deeper overhaul that cannot be accomplished when science is under the influence of capitalist logic and contingent on funding from businesses, whose ultimate bottom line is profitability, or business-friendly governments or non-governmental organisations, which depend ultimately on the same bottom line. What is needed is a scientific practice that is directly responsible (though not subordinated) to the rest of society at large and is especially responsive to oppressed peoples' initiatives and concerns. This would enable the development of scientific approaches amenable to the assembly of food systems compatible with ecosocialism. There are historical and current examples that demonstrate the prospective feasibility of such a change in the sciences, including agroecology in Cuba (see Chapter 6), the violently truncated re-direction of the biophysical sciences in the USSR of the 1920s (Gare 1993; Sheehan 1985), and the aforementioned work of *Science for the People* in the US (Schmalzer, Chard, and Botelho 2018).

However, food access for all will not be achieved even if these tough challenges are met—because they do not address the growing food distribution inequalities in society. Presently, there are arguments for an ecoSOCIALISM (as in social democratic parties) and for an ECOsocialism (as in green parties) when what is needed is a full-fledged ECOSOCIALISM. There are plenty of ways in which urban cultivation could play a role in promoting ecosocialist objectives of the sort outlined here. One is by rekindling integrative arrangements that typified urban planning in some Native American societies, as, for example, practised in Mayan cities, which, at one juncture, were organised around more self-sufficient urban farmsteads. Such planning mitigated the effects of extreme events as well as minimised long-distance food transport and energy necessary to accomplish it (Istendahl and Barthel 2018). Another possibility is linking ecosocially compatible urban cultivation movements across neighbourhoods, cities, and eventually countries, embracing similar movements in the countryside. If such rural movements do not exist, they can be created by squatting on farmland and putting it under ecologically sustainable communal use, as many who are in Brazil's Landless Workers' Movement have been doing successfully for decades (Hernandez 2020; Schwendler and Thompson 2017).

Instituting food justice (i.e. secure food access for all) and developing food sovereignty (i.e. food producers having decision-making power over food systems, especially in oppressed communities) is implicit in ecosocialist objectives. However, sovereignty is insufficient. This is because food production, distribution, and consumption need to be coordinated in ways that are mutually beneficial to all involved and that are ecologically sustainable (see also Cadieux and Slocum 2015; Jarosz 2014; Roman-Alcalá 2018; Tilzey 2018; Timmermann, Félix, and Tittonell 2018). Sovereignty is hardly achieved by producing what, in many cases, amounts to a small fraction of the food required for basic sustenance (García-Sempere et al. 2018). Even where enough food can be produced to fulfil subsistence needs, as in parts of Dar es Salaam, land tenure is insecure, and prevailing reliance on agrochemicals (self-poisoning, otherwise put) runs directly counter to any sovereignty potential. Moreover, food sovereignty at the level of a neighbourhood is grossly insufficient in the context of our capitalist global agri-food system, in which much everyday food consumption is predicated on inter-linkages across continents. In this conjuncture, sovereignty must begin to cut through global capitalist networks and coordinate changes across many areas of the world at once (see also McMichael 2015; Shattuck, Schiavoni, and van Gelder 2015)

The objectives of urban cultivation should therefore reach for much more than retaking the city. To be limited to cities is to risk a fate similar to that of the 1871 Commune, a revolutionary movement that took over Paris. Because of a lack of widespread social support elsewhere, joint Franco-Prussian ruling classes were able to squash it most brutally within a couple of months. In other words, there can be no ecosocialism or any radical break with capitalism in just one city or even in all cities combined. The effort needs to be international from the beginning, focused on a process of transforming cities into commons that span rural and urban areas, blurring for good the binary town and country separation. This is already being accomplished, but in destructive ways, in some countries by means of depopulating and industrialising the countryside. Ecosocialism is also a call, then, for an international socialisation of food systems, where the issue is not sovereignty per se but worldwide food distribution according to need and decided through nested participatory processes from the ground up. Hints of the effectiveness of this way of organising society already exist at regional levels, as exemplified by maroon-descendent communities in Venezuela and elsewhere (as well as the Chavista *comunas* movement), by the experiences of the Zapatista Army of National Liberation in Chiapas, México, and in some ways by what can be learned from the now ill-fated democratic confederalism of Rojava in northern Syria, and among some international Kurdish communities, despite decades of political persecution by various regional powers.

In conclusion, we find urban agriculture wanting when it comes to both ecological-environmental and social justice-equality concerns. Oft-assumed benefits like increasing sustainability and food justice do not stand up to empirical scrutiny. Technical solutions or recommendations that ride on such assumptions or that are implicitly capitalism-friendly are counter-productive, without overhauling the relationship between town and country. When not crassly biased towards profitability, food-growing technical expertise is politically suspect if it is easily diverted towards a further widening of social inequalities. Notwithstanding these faults, we deem urban food cultivation as important—if not crucial in many respects—in contributing to social sustainability and in reducing the distanciation between food production and consumption. In that way, cultivation counters some forms of capitalist alienation. Likewise, we see technical expertise as equally important in developing and spreading ecologically sustainable practices that simultaneously preserve human health. On this basis and as part of a wider set of strategies, urban food cultivation has a major role to play in envisioning and building ecologically sensible and egalitarian futures, that is, a post-capitalist ecosocialism.

References

Adamovsky, E., C. Albertani, B. Arditi, A.E. Ceceña, R. Gutiérrez, J. Holloway, F. López Bárcenas, G. López y Rivas, M. Modonesi, H. Ouviña, M. Thwaites Rey, S. Tischler, and R. Zibechi. 2011. *Pensar las Autonomías. Alternativas de Emancipación al Capital y el Estado*. [Thinking Autonomies. Alternatives of Emancipation to Capital and State] México: Sísifo Ediciones, Bajo Tierra.

Akuno, K. 2018. Casting Shadows. Chokwe Lumumba and the Struggle for Racial Justice and Economic Democracy in Jackson, Mississippi. In, *From the Streets to the State. Changing the World by Taking Power*, edited by P.C. Gray, 93–118. Albany: SUNY Press.

Akuno, K., A. Nangwaya, and Cooperation Jackson, eds. 2017. *Jackson Rising. The Struggle for Economic Democracy and Black Self-Determination in Jackson, Mississippi*. Jackson: Daraja Press.

Albo, G. 2007. The Limits of Eco-Localism: Scale, Strategy, Socialism. In, *Coming to Terms with Nature: Socialist Register 2007*, edited by L. Panitch, and C. Leys, 337–63. London: Merlin Press.

Baronnet, B., M. Mora Bayo, and R. Stahler-Sholk, eds. 2011. *Luchas "Muy Otras." Zapatismo y Autonomía en las Comunidades Indígenas de Chiapas*. ["Very Other" Struggles. Zapatismo and Autonomy in the Indigenous Communities of Chiapas] https://radiozapatista.org/pdf/libros/Luchas%20muy%20otras.pdf (Accessed 11 August 2020).

Barron, J. 2016. Community Gardening: Cultivating Subjectivities, Space, and Justice. *Local Environment* 22(9): 1142–58.

Engel-Di Mauro., S. 2017. Dialectics and Biophysical Worlds. *Science & Society* 81(3): 375–96.

Engel-Di Mauro., S. Forthcoming. Ecosocialism. In, *The Routledge Handbook of Development and Environment*, edited by B. McCusker, M. Ramutsindela, P. Solís, and W. Ahmed. New York: Routledge.

Forsyth, J. 1992. *A History of the Peoples of Siberia. Russia's North Asian Colony, 1581–1990.* Cambridge: Cambridge University Press.

García-Sempere, A., M. Hidalgo, H. Morales, B.G. Ferguson, A. Nazar-Beutelspacher, and P. Rosset. 2018. Urban Transition toward Food Sovereignty. *Globalizations* 15(3): 390–406.

Gare, A. 1993. Soviet Environmentalism: The Path not Taken. *Capitalism Nature Socialism* 4(3): 69–88.

Gare, A. 2016. *Philosophical Foundations of Ecological Civilization. A Manifesto for the Future.* New York: Routledge.

Garnett, T., and C. Godfray. 2012. *Sustainable Intensification in Agriculture: Navigating a Course Through Competing Food System Priorities.* Oxford: Oxford University Press.

Growing Communities. 2012. Growing Communities Food Zone: Towards a Sustainable and Resilient Food & Farming System. London Growing Communities. www.growingcommunities.org/ (Accessed 11 August 2020).

Haila, Y., and R. Levins. 1992. *Humanity and Nature. Ecology, Science and Society*. London: Pluto Press.

Haniotou, H., and E. Dalipi. 2018. Urban Gardens in Greece: A New Way of Living in the City. In, *The Urban Garden City. Cities and Nature,* edited by S. Glatron, and L. Granchamp, 245–68. Heidelberg: Springer Cham.

Hearn, M. 2010. *Common Ground in a Liquid City. Essays in Defense of an Urban Future.* Oakland: AK Press.

Hecht, D.K. 2012. How to Make a Villain: Rachel Carson and the Politics of Anti-Environ mentalism. *Endeavour* 36: 149–55.

Hernandez, A. 2020. The Emergence of Agroecology as a Political Tool in the Brazilian Landless Movement. *Local Environment* 25(3): 205–27.

Istendahl, C., and S. Barthel. 2018. Archaeology, History, and Urban Food Security. Integrating Cross-Cultural and Long-Term Perspectives. In, *Routledge Handbook of Landscape and* Food, edited by J. Zeunert, and T. Waterman, 61–72. London: Routledge.

Jarosz, L. 2014. Comparing Food Security and Food Sovereignty Discourses. *Dialogues in Human Geography* 4(2): 168–81.

Kovel, J. 1991. *History and Spirit. An Inquiry into the Philosophy of Liberation.* Boston, MA: Beacon Press.

Kovel, J. 1995. Ecological Marxism and Dialectic. *Capitalism Nature Socialism* 6(4): 31–50.

Kovel, J. 1998. Dialectic as Praxis. *Science & Society* 62(1): 474–82.

Kovel, J. 2005. The Ecofeminist Ground of Ecosocialism. *Capitalism Nature Socialism* 16(2): 2–8.

Kovel, J. 2007. *The Enemy of Nature. The End of Capitalism or the End of the World*? 2nd ed. London: Zed Books.

Kovel, J. 2014. The Future Will Be Ecosocialist, Because Without Ecosocialism, There Will Be No Future. In, *Imagine Living in a Socialist USA*, edited by F. Goldin, D. Smith, and M.S. Smith, 25–32. New York: Harper Perennial.

Kovel, J., and M. Löwy. 2001. An Ecosocialist Manifesto. http://ecosocialisthorizons.com/2001/09/an-ecosocialist-manifesto/ (Accessed 11 August 2020).

Kummu, M., H. de Moel, M. Porkka, S. Siebert, O. Varis, and P. Ward. 2012. Lost Food, Wasted Resources: Global Food Supply Chain Losses and Their Impacts on Freshwater, Cropland, and Fertilizer Use. *The Science of the Total Environment* 438(1): 477–89.

Bedford, D., and D. Irving-Stephens. 2000. *The Tragedy of Progress: Marxism, Modernity, and the Aboriginal Question*. Halifax: Fernwood.

Bellwood-Howard, I., and E. Bogweh Nchanji. 2017. The Marketing of Vegetables in a Northern Ghanaian City: Implications and Trajectories. In, *Global Urban Agriculture: Convergence of Theory and Practice Between North and South*, edited by A.M.G.A. Winkler-Prins, 79–92. Boston, MA: CABI.

Borch, C., and M. Kornberger, eds. 2015. *Urban Commons. Rethinking the City*. Abingdon: Routledge.

Born, B., and M. Purcell. 2006. Avoiding the Local Trap: Scale and Food Systems in Planning Research. *Journal of Planning Education and Research* 26: 195–207.

Boukharaeva, L., and M. Marloie, eds. 2015. *Family Urban Agriculture in Russia. Lessons and Prospects*. Cham: Springer.

Buchanan, L., Q. Bui, and J.K. Patel. 2020. Black Lives Matter May Be the Largest Movement in US History. *The New York Times*, 3 July. www.nytimes.com/interactive/2020/07/03/ (Accessed 11 August 20202).

Busà, A. 2017. *The Creative Destruction of New York City: Engineering the City for the Elite*. New York: Oxford University Press.

Cadieux, K.V., and R. Slocum. 2015. What Does It Mean to *Do* Food Justice? *Journal of Political Ecology* 22(1). http://jpe.library.arizona.edu/volume_22/Cadieuxslocum.pdf (Accessed 11 August 2020).

Cassidy, E., P. West, J. Gerber, and J. Foley. 2013. Redefining Agricultural Yields: From Tonnes to People Nourished Per Hectare. *Environmental Research Letters* 8: 1–8.

CDC. 2020. *Global COVID-19*. Washington, DC: Centers for Disease Control and Prevention, US Department of Health and Human Services. www.cdc.gov/.

Chotiner, Isaac. 2020. A Black Lives Matter Co-Founder Explains Why This Time Is Different. *The New Yorker*, June 3. www.newyorker.com/news/q-and-a/black-lives-matter/ (Accessed 11 August 2020).

Clark, J. 2014. It Is What It Isn't! A Defence of Dialectic. *Review* 31. http://review31.co.uk/essay/view/7/it-is-what-it-isn%27t-a-defence-of-the-dialectic (Accessed 11 August 2020).

Clark, J. 2016. *The Tragedy of Common Sense*. Regina: Changing Suns Press.

Colding, J., S. Barthel, P. Bendt, R. Snep, W. van der Knaap, and H. Ernstson. 2013. Urban Green Commons: Insights on Urban Common Property Systems. *Global Environmental Change* 23(5): 1039–51.

Dellenbaugh, M., M. Kip, M. Bieniok, A.K. Müller, and M. Schwegmann, eds. 2015. *Urban Commons Moving Beyond State and Market*. Berlin: Bauverlag.

DenHoed, Andrea. 2016. The Forgotten Lessons of the American Eugenics Movement. *The New Yorker*, 27 April. www.newyorker.com/books/page-turner/the-forgotten-lessons-/ (Accessed 11 August 2020).

Dickens, P. 1992. *Society and Nature. Towards a Green Social Theory*. Philadelphia, PA: Temple University Press.

Eidelman, T.A., and S. Safransky. 2020. The Urban Commons: A Keyword Essay. *Urban Geography*, 20 May. https://doi.org/10.1080/02723638.2020.1742466 (Accessed 11 August 2020).

Ela, N. 2006. Urban Commons as Property Experiment: Mapping Chicago's Farms and Gardens. *Fordham Urban Law Journal* 43: 247–94.

Elflein, J. 2020. COVID-19 Deaths Worldwide as of August 10, 2020, by Country. *Statista*. www.statista.com/ (Accessed 11 August 2020).

Engel-Di Mauro, S. 2014. *Ecology, Soils, and the Left: An Eco-Social Approach*. New York: Palgrave Macmillan.

Lemarchand, F. 2018. Russian Collective Gardens: A Story of Institution and Remembrance. In, *The Urban Garden City. Cities and Nature*, edited by S. Glatron, and L. Granchamp, 79–97. Heidelberg: Springer Cham.

Levins, R., and R. Lewontin. 1985. *The Dialectical Biologist*. Cambridge: Harvard University Press.

Levins, R., and R. Lewontin. 2007. *Biology Under the Influence: Dialectical Essays on Ecology, Agriculture, and Health*. New York: Monthly Review Press.

Löwy, M. 2005. What Is Ecosocialism? *Capitalism Nature Socialism* 16(2): 15–24.

Löwy, M. 2011. *Écosocialisme: L'Alternative Radicale à la Catastrophe Écologique Capitaliste*. Paris: Fayard.

Macdiarmid, J., J. Kyle, G. Horgan, and G. Livewell. 2014. A Balance of Healthy and Sustainable Food Choices. https://livewellforlife.eu/wp-content/uploads/2013/02/A-balance-of-healthy-and-sustainable-food-choices.pdf (Accessed 11 August 2020).

Mark, J. 2020. The Pandemic Is Reviving the Push for Locally Produced Foods. *Sierra Club*, 15 April. www.sierraclub.org/sierra/rebirth-food-sovereignty-movement (Accessed 11 August 2020).

Martin, G., R. Clift, and I. Christie. 2016. Urban Cultivation and Its Contributions to Sustainability: Nibbles of Food but Oodles of Social Capital. *Sustainability* 8(5): 409.

Martínez López, M.A. 2018. *The Urban Politics of Squatters Movements*. New York: Palgrave Macmillan.

Marx, K. 1959 [1853]. The British Rule of India. In, *On Colonialism*, edited by K. Marx, and F. Engels, 32–9. Moscow: Foreign Languages Publishing House.

Marx, K. 1978 [1844)]. Economic and Philosophic Manuscripts of 1844. In, *The Marx-Engels Reader*, edited by Robert C. Tucker, 2nd ed., 66–125. New York: W.W. Norton & Company.

Marx, K. 1992 [1867)]. *Capital. A Critical Analysis of Capitalist Production. Volume 1*. Translated by S. Moore, and E. Aveling. New York: International Publishers.

Massachusetts Jobs with Justice. 2020. Covid-19 Mutual Aid Networks. www.massjwj.net/news/2020/3/17/cover-19-mutual-aid-networks (Accessed 30 April 2021).

McMichael, P. 2015. The Land Question in the Food Sovereignty Project. *Globalizations* 12(4): 434–51.

Mies, M., and V. Bennholdt-Thomsen. 1999. *The Subsistence Perspective: Beyond the Globalized Economy*. London: Zed Books.

Mies, M., and V. Shiva. 1993. *Ecofeminism*. London: Zed.

Mies, M., and V. Shiva. 1993. *Ecofeminism*. London: Zed.

Mudu, P., and A. Marini. 2018. Radical Urban Horticulture for Food Autonomy: Beyond the Community Gardens Experience. *Antipode* 50(2): 549–73.

O'Connor, J. 1988. Capitalism, Nature, Socialism. A Theoretical Introduction. *Capitalism Nature Socialism* 1(1): 11–38.

Paolucci, P. 2007. *Marx's Scientific Dialectics: A Methodological Treatise for a New Century*. Boston, MA: Brill Academic Publishers.

Pepper, D. 2003. *Eco-Socialism: From Deep Ecology to Social Justice*, 2nd ed. London: Routledge.

Roberts, N. 2015. *Freedom as Marronage*. Chicago: The University of Chicago Press.

Roman-Alcalá, A. 2018. (Relative) Autonomism, Policy Currents and the Politics of Mobilisation for Food Sovereignty in the United States: The Case of Occupy the Farm. *Local Environment* 23(6): 619–34.

Sage, C. 2012. Addressing the Faustian Bargain of the Modern Food System: Connecting Sustainable Agriculture with Sustainable Consumption. *International Journal of Agricultural Sustainability* 10: 204–7.

Salleh, A. 1997. *Ecofeminism as Politics. Nature, Marx, and the Postmodern*. London: Zed Books.

Salleh, A., ed. 2009. *Eco-Sufficiency and Global Justice: Women Write Political Ecology*. London: Pluto Press.

Schmalzer, S., D.S. Chard, and A. Botelho, eds. 2018. *Science for the People: Documents from America's Movement of Radical Scientists*. Amherst: University of Massachusetts Press.

Schwartzman, D. 1996. Solar Communism. *Science & Society* 60(3): 307–31.

Schwartzman, D. 2009. Ecosocialism or Ecocatastrophe? *Capitalism Nature Socialism* 20(1): 6–33.

Schwendler, S.F., and L.A. Thompson. 2017. An Education in Gender and Agroecology in Brazil's Landless Rural Workers' Movement. *Gender and Education* 29(1): 100–14.

Scott, A. 2020. A Comeback for Victory Gardens. *San Francisco Chronicle*, 5 April: J6.

Sekulich, T. 2019. Top Ten Agribusinesses in the World. *Tharawat Magazine*, 7 February. www.tharawat-magazine.com/facts/top-ten-agribusiness-companies/ (Accessed 11 August 2020).

Selden, T.M., and T.A. Berdahl. 2020. COVID-19 and Racial/Ethnic Disparities in Health Risk, Employment, and Household Composition. *Human Affairs*, 20 July. https://europepmc.org/article/med/32663045 (Accessed 11 August 2020).

Sharzer, G. 2012. A Critique of Localist Political Economy and Urban Agriculture. *Historical Materialism* 20(4): 75–114.

Shattuck, A., C.M. Schiavoni, and Z. VanGelder. 2015. Translating the Politics of Food Sovereignty: Digging into Contradictions, Uncovering New Dimensions. *Globalizations* 12(4): 421–33.

Sheehan, H. 1985. *Marxism and the Philosophy of Science: A Critical History*. Atlantic Highlands: Humanities Press.

Shoatz, R.M. 2013. *Maroon the Implacable: The Collected Writings of Russell Maroon Shoatz*. Edited by Q. Saul, and F. Ho. Oakland: PM Press.

Siebert, A. 2020. Transforming Urban Food Systems in South Africa: Unfolding Food Sovereignty in the City. *The Journal of Peasant Studies* 47(2): 401–19.

Smith, A., T. Hargreaves, S. Hielscher, M. Martiskainen, and G. Seyfang. 2016. Making the Most of Community Energies: Three Perspectives on Grassroots Innovation. *Environment and Planning A* 48(2): 407–32.

Subcomandante Marcos, and Y. Le Bot. 1997. *El Sueño Zapatista*. [The Zapatista Dream] https://enriquedussel.com/txt/Textos_200_Obras/PyF_revolucionarios_marxistas/Sueno_zapatista-Yvon_Le_Bot.pdf (Accessed 11 August 20202).

Thompson, S. 2013. *Live Well for LIFE: A Balance of Healthy and Sustainable Food Choices for France, Spain, and Sweden*. Woking: World Wide Fund for Nature. http://livewellforlife.eu/wp-content/uploads/2013/02/A-balance-of-healthy-and-sustainable-food-choices.pdf (Accessed 11 August 20202).

Tilzey, M. 2018. *Political Ecology, Food Regimes, and Food Sovereignty. Crisis, Resistance, and Resilience*. New York: Palgrave Macmillan.

Timmermann, C., G.F. Félix, and P. Tittonell. 2018. Food Sovereignty and Consumer Sovereignty: Two Antagonistic Goals? *Agroecology & Sustainable Food Systems* 42(3): 274–98.

Tornaghi, C. 2017. Urban Agriculture in the Food-Disabling City: (Re)defining Urban Food Justice, Reimagining a Politics of Empowerment. *Antipode* 49(3): 781–801.

Turner, T., and L. Brownhill. 2004. We Want Our Land back: Gendered Class Analysis, the Second Contradiction of Capitalism and Social Movement Theory. *Capitalism Nature Socialism* 15(4): 21–40.

UN. 2019. *World Urbanization Prospects 2018*. New York: Department of Economic and Social Affairs, United Nations.

Wallerstein, I. 1999. Ecology and Capitalist Costs of Production: No Exit. In, *Ecology and the World-System*, edited by W.L. Goldfrank, D. Goodman, and A. Szasz, 3–11. Westport: Greenwood Press.

Wan, P.Y.-Z. 2012. Dialectics, Complexity, and the Systemic Approach: Toward a Critical Reconciliation. *Philosophy of the Social Sciences* 43(4): 411–52.

Weinberger, H. 2020. WWII-Era 'Victory Gardens' Make a Comeback Amid Coronavirus. *Crosscut*, 27 March. https://crosscut.com/2020/23/wwii-era-victory-gardens (Accessed 11 August 20202).

Wharton, R. 2020. Community Gardens Learn to Adapt. *The New York Times*, 15 April 15: D6.

WHO. 2020. *Rolling Updates on Coronavirus Disease (COVID-19)*. Geneva: World Health Organization, United Nations. www.who.int/ (Accessed 11 August 2020).

World Bank. 2020. *GDP per Capita, PPP (Current International $)*. Washington DC. https://data.worldbank.org/indicator/NY.GDP.PCAP.PP.CD (Accessed 11 August 2020).

INDEX